# AQA GCSE
# MATHS

## Higher

## COMPLETE REVISION AND PRACTICE

Naomi Bartholomew-Millar

Paul Hunt

Victoria Trumper

OXFORD
UNIVERSITY PRESS

# Contents

 Shade in each level of the circle as you feel more confident and ready for your exam.

# How to use this book

This book uses a three-step approach to revision: **Knowledge**, **Retrieval**, and **Practice**. It is important that you do all three; they work together to make your revision effective.

 **Knowledge**

**Knowledge** comes first. Each chapter starts with a **Knowledge Organiser**. These are clear easy-to-understand, concise summaries of the content that you need to know for your exam. The information is organised to show how one idea flows into the next so you can learn how everything is tied together, rather than lots of disconnected facts.

**Worked example**

**Worked examples** offer step by step guidance on working through a question, to a solution.

**LINK**

The **Link** box highlights a reference to a related topic you may want to refer to.

**REVISION TIP**

**Revision tips** offer you helpful advice and guidance to aid your revision and help you to understand key concepts and remember them.

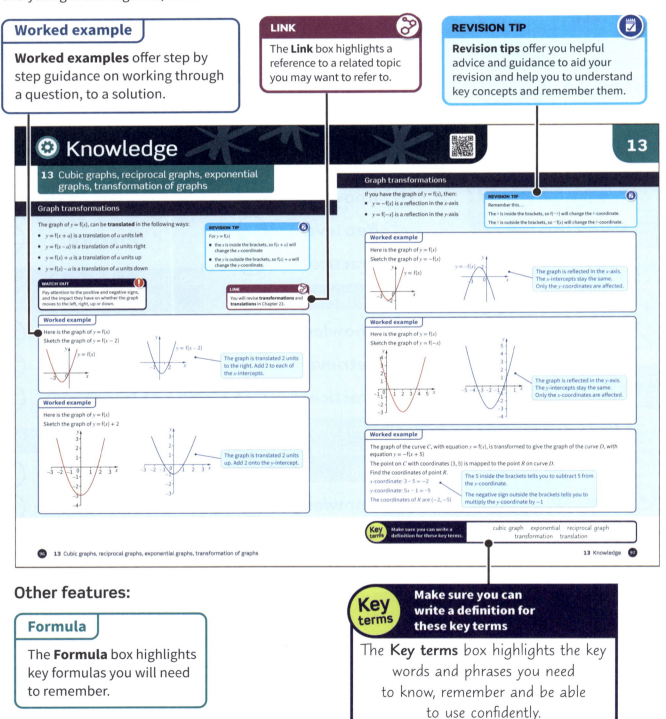

## Other features:

**Formula**

The **Formula** box highlights key formulas you will need to remember.

**Key terms** Make sure you can write a definition for these key terms

The **Key terms** box highlights the key words and phrases you need to know, remember and be able to use confidently.

## Retrieval

The **Retrieval questions** help you learn and quickly recall the information you've acquired. These are short questions and answers about the content in the Knowledge Organiser you have just reviewed. Cover up the answers with some paper and write down as many answers as you can from memory. Check back to the Knowledge Organiser for any you got wrong, then cover the answers and attempt all the questions again until you can answer *all* the questions correctly.

Make sure you revisit the retrieval questions on different days to help them stick in your memory. You need to write down the answers each time, or say them out loud, otherwise it won't work.

### Previous Questions

Each chapter also has some **Retrieval questions** from **previous chapters**. Answer these to see if you can remember the content from the earlier chapters. If you get the answers wrong, go back and do the Retrieval questions for the earlier chapters again.

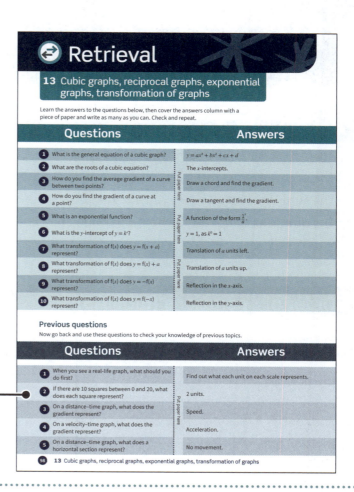

## Retrieval

**13** Cubic graphs, reciprocal graphs, exponential graphs, transformation of graphs

Learn the answers to the questions below, then cover the answers column with a piece of paper and write as many as you can. Check and repeat.

| Questions | Answers |
|---|---|
| 1 What is the general equation of a cubic graph? | $y = ax^3 + bx^2 + cx + d$ |
| 2 What are the roots of a cubic equation? | The $x$-intercepts. |
| 3 How do you find the average gradient of a curve between two points? | Draw a chord and find the gradient. |
| 4 How do you find the gradient of a curve at a point? | Draw a tangent and find the gradient. |
| 5 What is an exponential function? | A function of the form $\frac{1}{a}$. |
| 6 What is the $y$-intercept of $y = k^x$? | $y = 1$, as $k^0 = 1$ |
| 7 What transformation of f($x$) does $y = $ f($x + a$) represent? | Translation of $a$ units left. |
| 8 What transformation of f($x$) does $y = $ f($x$) $+ a$ represent? | Translation of $a$ units up. |
| 9 What transformation of f($x$) does $y = -$f($x$) represent? | Reflection in the $x$-axis. |
| 10 What transformation of f($x$) does $y = $ f($-x$) represent? | Reflection in the $y$-axis. |

**Previous questions**

Now go back and use these questions to check your knowledge of previous topics.

| Questions | Answers |
|---|---|
| 1 When you see a real-life graph, what should you do first? | Find out what each unit on each scale represents. |
| 2 If there are 10 squares between 0 and 20, what does each square represent? | 2 units. |
| 3 On a distance–time graph, what does the gradient represent? | Speed. |
| 4 On a velocity–time graph, what does the gradient represent? | Acceleration. |
| 5 On a distance–time graph, what does a horizontal section represent? | No movement. |

**98** **13** Cubic graphs, reciprocal graphs, exponential graphs, transformation of graphs

## Practice

Once you think you know the Knowledge Organiser and Retrieval answers really well, you can move on to the final stage: **Practice**.

Each chapter has **exam-style questions**, including some questions from previous chapters, to help you apply all the knowledge you have learnt and can retrieve.

### Answers and Glossary

You can scan the QR code at any time to access the sample answers and mark schemes for all the exam-style questions, glossary containing definitions of the key terms, as well as further revision support. Visit go.oup.com/OR/GCSE/A/Maths/H

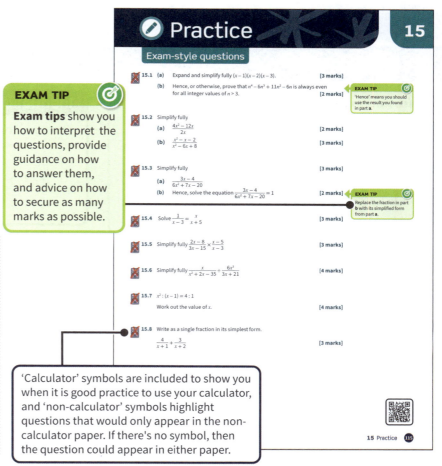

## Practice                                                                15

**Exam-style questions**

**15.1 (a)** Expand and simplify fully $(x - 1)(x - 2)(x - 3)$.  **[3 marks]**
**(b)** Hence, or otherwise, prove that $n^4 - 6n^3 + 11n^2 - 6n$ is always even for all integer values of $n > 3$.  **[2 marks]**

> **EXAM TIP**
> 'Hence' means you should use the result you found in part a.

**15.2** Simplify fully
**(a)** $\frac{4x^2 - 12x}{2x}$  **[2 marks]**
**(b)** $\frac{x^2 - x - 2}{x^2 - 6x + 8}$  **[3 marks]**

**15.3** Simplify fully  **[3 marks]**
**(a)** $\frac{3x - 4}{6x^2 + 7x - 20}$
**(b)** Hence, solve the equation $\frac{3x - 4}{6x^2 + 7x - 20} = 1$  **[2 marks]**

> **EXAM TIP**
> Replace the fraction in part b with its simplified form from part a.

**15.4** Solve $\frac{1}{x - 3} = \frac{x}{x + 5}$  **[3 marks]**

**15.5** Simplify fully $\frac{2x - 8}{3x - 15} \times \frac{x - 5}{x - 3}$  **[3 marks]**

**15.6** Simplify fully $\frac{x}{x^2 + 2x - 35} \div \frac{6x^3}{3x + 21}$  **[4 marks]**

**15.7** $x^2 : (x - 1) = 4 : 1$
Work out the value of $x$.  **[4 marks]**

**15.8** Write as a single fraction in its simplest form.
$\frac{4}{x + 1} + \frac{3}{x + 2}$  **[3 marks]**

**15** Practice **115**

> **EXAM TIP**
> **Exam tips** show you how to interpret the questions, provide guidance on how to answer them, and advice on how to secure as many marks as possible.

'Calculator' symbols are included to show you when it is good practice to use your calculator, and 'non-calculator' symbols highlight questions that would only appear in the non-calculator paper. If there's no symbol, then the question could appear in either paper.

v

# ⚙ Knowledge

## 1 Calculating with all four operations, place value, powers and indices

### Definitions

An **integer** is a whole number, such as 1, 73, 946, −17.

The **product** is the result when you multiply numbers.

The **sum** is the result when you add numbers.

The **quotient** is the result when you divide one number by another.

**Commutative** means that the calculation can be done in any order. For example, addition and multiplication are commutative:

$$7 + 4 = 4 + 7 \qquad 8 \times 2 = 2 \times 8$$

But subtraction and division are not:

$$7 - 4 \neq 4 - 7 \qquad 8 \div 2 \neq 2 \div 8$$

### Order of operations (BIDMAS)

Use BIDMAS to help you do calculations in the correct order.

**B**rackets ( )

**I**ndices (powers and roots)

**D**ivision and **M**ultplication

**A**ddition and **S**ubtraction

**B**

**I**

**D M**

**A S**

**Worked example**

1. $43 - 9 \times 4 + 5$

   $43 - 9 \times 4 + 5$ — BIDMAS Multiplication first.

   $= 43 - 36 + 5$ — BIDMAS Addition and Subtraction.

   $= 7 + 5 = 12$

2. $32 - 6^2 \div 2$

   $32 - 6^2 \div 2$ — BIDMAS Indices (powers) first.

   $= 32 - 36 \div 2$ — BIDMAS Division.

   $= 32 - 18 = 14$

### Ordering numbers

**Worked example**

Which is larger, 1.3 or 1.105?

   1 . 3 **0 0**

   1 . 1 0 5

   1.3 > 1.105

Write in columns, then write in 0s so both numbers have the same number of decimal places.

**Inequality symbols**

The small end points to the small number.

## Negative numbers

**Rules for adding and subtracting negative numbers**

Adding a positive number.

$5 + 2$ $+2$

Adding a negative number is the same as subtracting.

$5 + -2$ $-2$

$5 + -2$
replace $+ -$ with $-$
$= 5 - 2 = 3$

Subtracting a positive number.

$-4 - 3$ $-3$

Subtracting a negative number is the same as adding.

$-4 - -3$ $+3$

$-4 - -3$
replace $- -$ with $+$
$-4 + 3 = -1$

**Worked example**

$1 - 3 - -6$

$1 - 3 \underline{- -} 6$ — Replace $- -$ with $+$

$= 1 - 3 + 6$

$= -2 + 6$

$= 4$

**Rules for multiplying and dividing negative numbers**

positive $\times$ or $\div$ positive = positive
positive $\times$ or $\div$ negative = negative
negative $\times$ or $\div$ positive = negative
negative $\times$ or $\div$ negative = positive

**Worked example**

1. $4 \times (-2) = -8$ — positive $\times$ negative = negative

2. $(-3) \times 5 = -15$ — negative $\times$ positive = negative

3. $(-3) \times (-4) = 12$ — negative $\times$ negative = positive

4. $6 \div (-2) = -3$ — positive $\div$ negative = negative

5. $(-8) \div (-2) = 4$ — negative $\div$ negative = positive

## Calculating with decimals

When adding or subtracting decimals:

- Align the decimal points and carry out the calculation as usual
- Make sure to include the decimal point in your answer.

When multiplying or dividing decimals:

- Ignore the decimals and carry out the calculation as usual
- Then work out where to place the decimal.

# ⚙ Knowledge

## 1 Calculating with all four operations, place value, powers and indices

---

## Calculating with decimals

### Worked example

Work out

**1.** $21.47 + 32.85$

$$\begin{array}{r} 2\,1\,.\,4\,7 \\ +\ 3\,2\,.\,8\,5 \\ \hline 5\,4\,.\,3\,2 \\ {\scriptstyle 1\ \ \ 1} \end{array}$$

Align decimal points.

Add numbers vertically. Start with the left-hand column, and move right.

Make sure you carry the tens correctly.

**2.** $35.7 - 26.34$

$$\begin{array}{r} {}^{2}\cancel{3}\,{}^{1}5\,.\,{}^{6}\cancel{7}\,{}^{1}0 \\ -\ 2\,6\,.\,3\,4 \\ \hline 9\,.\,3\,6 \end{array}$$

Align decimal points.

Add a zero so you have the same number of place values in each number.

Subtract the second row from the top row. Start with the left-hand column, and move right.

Make a place value exchange when you need to.

**3.** $8.16 \times 4.9$

$$\begin{array}{r} 8\,1\,6 \\ \times\ \ \ 4\,9 \\ \hline 7\,3\,4\,4 \\ {\scriptstyle 1\ 5} \\ 3\,2\,6\,4\,0 \\ {\scriptstyle 2} \\ \hline 3\,9\,9\,8\,4 \end{array}$$

$8.16 = 816 \div 100$

$4.9 = 49 \div 10$

$39\,984 \div 100 \div 10$
$= 39.984$

Ignore decimal points.

Multiply each digit of the top number by '9'.

Multiply each digit of the top number by '40'.

Add the products of the two multiplications together.

Work out what power of 10 to divide by to find the answer to the original question.

**4.** $11.07 \div 0.9$

$$\frac{11.07}{0.9} = \frac{110.7}{9}$$

(× 10 above and below)

$$9)\overline{11\,{}^{2}0.\,{}^{2}7} = 12.3$$

Write the division as a fraction and find an equivalent with an integer denominator.

Carry out the new division working from left to right.

---

## Bases and indices

$7^{5}$ — index

— base

$a^0 = 1$

Example:

$7^0 = 1$

Any number (or letter) raised to the power of 0 is 1.

$a^n \times a^m = a^{n+m}$

Example:

$3^2 \times 3^4 = 3^{2+4} = 3^6$

To multiply numbers with the same base, add the powers.

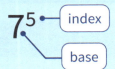

## Bases and indices

| $a^n \div a^m = a^{n-m}$ | $\left(a^n\right)^m = a^{n \times m}$ | $a^{-1} = \dfrac{1}{a}$ | $a^{\frac{1}{n}} = \sqrt[n]{a}$ |
|---|---|---|---|
| Example:<br>$5^7 \div 5^3 = 5^{7-3} = 5^4$ | Example:<br>$(4^3)^5 = 4^{3 \times 5} = 4^{15}$ | Example:<br>$2^{-1} = \dfrac{1}{2}$ | Example:<br>$8^{\frac{1}{3}} = \sqrt[3]{8}$ |
| To divide numbers with the same base, subtract the powers. | To raise a power by another power, multiply the powers. | A number raised to the power of –1 gives the reciprocal of the number. | A power of $\frac{1}{n}$ gives the $n$th root. |

An indices question will either ask you to calculate or simplify.

- **Calculate** (or work out) means find the numerical answer.

- **Simplify** means write in a simpler form, but your answer will still contain powers.

### Worked example

Work out the value of:

**(a)** $\left(\dfrac{9}{4}\right)^{\frac{1}{2}}$

$\left(\dfrac{9}{4}\right)^{\frac{1}{2}} = \sqrt{\dfrac{9}{4}}$

$= \dfrac{\sqrt{9}}{\sqrt{4}}$

$= \dfrac{3}{2}$

> Use $a^{\frac{1}{n}} = \sqrt[n]{a}$
>
> Apply the power to the numerator and denominator separately.

**(b)** $64^{-\frac{2}{3}}$

$64^{-\frac{2}{3}} = \dfrac{1}{64^{\frac{2}{3}}}$

$= \dfrac{1}{\left(64^{\frac{1}{3}}\right)^2}$

$= \dfrac{1}{\left(\sqrt[3]{64}\right)^2}$

$= \dfrac{1}{(4)^2}$

$= \dfrac{1}{16}$

> Use $a^{-1} = \dfrac{1}{a}$
>
> Use $(a^n)^m = a^{n \times m}$
>
> Use $a^{\frac{1}{n}} = \sqrt[n]{a}$

**(c)** Simplify $\sqrt{\dfrac{5^3}{5^{-5}}}$

Give your answer as a power of 5.

$\sqrt{\dfrac{5^3}{5^{-5}}} = \left(\dfrac{5^3}{5^{-5}}\right)^{\frac{1}{2}}$

$= \left(5^{3-(-5)}\right)^{\frac{1}{2}}$

$= \left(5^8\right)^{\frac{1}{2}}$

$= 5^{8 \times \frac{1}{2}}$

$= 5^4$

> Use $a^{\frac{1}{n}} = \sqrt[n]{a}$
>
> Use $a^n \div a^m = a^{n-m}$
>
> $(a^n)^m = a^{n \times m}$

## Indices on your calculator

The symbols on different calculators vary slightly, but often look like this:

Cube root:

Square root:

Cube:

Square: $x^2$

Any root:

Any power:

## 1 Calculating with all four operations, place value, powers and indices

### Listing outcomes

A list or table showing all possible outcomes of an experiment or of **combined experiments** is called the **possibility space** or **sample space**. For a combined experiment:

*number of possible outcomes* $= m \times n$

Where

- $m$ = possible outcomes of one experiment

- $n$ = possible outcomes of the other experiment

> **Worked example**
>
> This padlock has digits 0–9 on each dial.
>
> How many possible combinations are there?
>
>
>
> number of possible combinations =
>
> number of possible digits on dial 1 multiplied by number of possible digits on dial 2 multiplied by number of possible digits on dial 3.
>
> There are 10 possible digits for each of the three dials.
> Total number of possible combinations $= 10 \times 10 \times 10$
> $$= 1000$$

**Make sure you can write a definition for these key terms.**

calculate   commutative   combined experiment   cube
possibility space   power   product   quotient
sample space   simplify   sum

## 1 Calculating with all four operations, place value, powers and indices

Learn the answers to the questions below, then cover the answers column with a piece of paper and write as many as you can. Check and repeat.

## Questions | Answers

| # | Question | Answer |
|---|----------|--------|
| 1 | What is an integer? | A whole number. |
| 2 | What does commutative mean? | Can be done in any order. |
| 3 | What do the letters of BIDMAS stand for? | Brackets, indices, division, multiplication, addition and subtraction. |
| 4 | What does BIDMAS tell you? | The order in which to do a calculation. |
| 5 | When you multiply a negative number and a positive number, what sign does the answer have? | Negative. |
| 6 | When you divide a negative number by a negative number, what sign does the answer have? | Positive. |
| 7 | What is the product of two or more numbers | The result you get when you multply them together. |
| 8 | What is the sum of two or more numbers? | The result you get when you add them together. |
| 9 | What is the quotient of two numbers? | The result you get when you divide one of the numbers by the other. |
| 10 | What is a list or table showing all possible outcomes of an experiment called | A possibility space or sample space. |
| 11 | What is the value of any number or variable raised to the power of 0? | 1 |
| 12 | Write examples to show each of the six laws of indices. | $5^0 = 1$<br>$5^3 \times 5^5 = 5^{3+5} = 5^8$<br>$5^8 \div 5^5 = 5^{8-5} = 5^3$<br>$(5^5)^3 = 5^{5 \times 3} = 5^{15}$<br>$5^{-1} = \frac{1}{5}$<br>$5^{\frac{1}{2}} = \sqrt{5}$<br>(any base number can be used) |
| 13 | Which buttons on your calculator would you use to find $4^6$? | $4 \boxed{x^n} 6 =$ |
| 14 | What is $27^{\frac{2}{3}}$ the same as calculating? | $(\sqrt[3]{27})^2$ |
| 15 | Raising a number to which power gives you the reciprocal of a number? | $-1$ |

*Put paper here*

# Practice

**1.1** Work out

    **(a)**    $25.043 - 17.82$            **[2 marks]**

    **(b)**    $17.12 \div 0.8$            **[2 marks]**

**1.2** Work out

    **(a)**    $\dfrac{4 \times 5^2}{4 \times 5 \div 2}$            **[1 mark]**

    **(b)**    $(1 - 0.1) \times 4 - (-10)$            **[2 marks]**

    **(c)**    $\dfrac{(-0.2) \times (-6)}{-1 + 0.7}$            **[2 marks]**

**1.3** Marina's fence measures 1.4 m by 10.5 m.
It costs £0.60 per square metre to paint the fence.
How much does it cost to paint the fence in total?    **[3 marks]**

**1.4** Thema says that multiplying by 0.01 is the same as dividing by 100.
Is Thema correct? Give a reason for your answer.    **[1 mark]**

**1.5** Peter says that $2^3 \times 5^2$ simplifies to $10^5$. Peter is wrong.
Explain why.    **[1 mark]**

 **1.6** Simplify fully $\dfrac{(2^7) \times (2^4)^{-1}}{2}$

Give your answer as a power of 2    **[2 marks]**

> **EXAM TIP**
>
> An index of $-1$ is a reciprocal. An index of $\frac{1}{n}$ is a root.

**1.7** Work out the value of

    **(a)**    $\left(3^{\frac{1}{4}}\right)^{\frac{1}{4}}$            **[1 mark]**

    **(b)**    $\dfrac{a^{\frac{2}{3}}}{\sqrt{a}}$            **[1 mark]**

**1.8** Work out the value of

    **(a)**    $8^{\frac{2}{3}}$            **[2 marks]**

    **(b)**    $\left(\dfrac{16}{9}\right)^{-\frac{2}{4}}$            **[3 marks]**

**1.9** $3 \times \sqrt{27} = 3^n$

Work out the value of $n$. **[3 marks]**

 **1.10** Find the value of the reciprocal of $2\frac{2}{3}$

Write your answer as a decimal. **[2 marks]**

 **1.11** $\frac{1}{a} = 2^{-3}$   $2^b = 2\sqrt{2}$   $\frac{1}{\sqrt[c]{2}} = 2^{-\frac{1}{3}}$

Work out the value of $\frac{ab}{c}$. **[4 marks]**

 **1.12** Put these numbers in ascending order. **[2 marks]**

$\left(-\frac{1}{3}\right)^3, 3^{-2}, \left(-\frac{1}{2}\right)^4, \left(-\frac{1}{4}\right)^{-2}, 2^{-1}$

 **1.13** A four-digit pin code can only use the digits 1 to 9.
Digits can be used more than once.

**(a)** How many possible pin codes are there? **[2 marks]**

**(b)** How many of the pin codes use the digit 8 at least once? **[2 marks]**

**1.14** A shop sells:

15 types of chocolate bar

6 flavours of crisps

8 choices of canned drink.

Isabella wants to choose
a chocolate bar and a canned drink
**or** a packet of crisps and a canned drink
**or** a chocolate bar, a packet of crisps and a canned drink.

Show that there are 888 different combinations for Isabella to
choose from. **[4 marks]**

**1.15** In a games competition, there are eight teams.
Each team plays four games against each of the other teams.

Work out how many games are played in total. **[2 marks]**

# Knowledge

## 2 Rounding, truncating, error intervals, and estimating

### Rounding

When numbers are **rounded** they are written only to a given **place value**.

**Worked example**

Round 15 726 to the nearest 1000.

1 <u>5</u> 7 2 6 — Underline the 1000s **digit**.

1 5 ⑦ 2 6 — Look at the next digit. 5 or more, so round up.

1 6 0 0 0 — Add 1 to round up. — Insert three 0s to keep the place value (1000s).

15 000     15 726   16 000

**Worked example**

Round 0.143 76 to 2 decimal places.

0 . 1 <u>4</u> 3 7 6 — Underline the digit in the 2nd decimal place (100ths digit).

0 . 1 4 ③ 7 6 — Look at the next digit. Less than 5, so round down.

0 . 1 4 (2 d.p.) — State the level of accuracy.

Delete the digits after the 2nd decimal place. Don't insert any 0s, as 0.14 has correct place value (100ths).

**Significant figures (s.f.)** are digits in a number that contribute to its accuracy.

1st s.f.  2nd s.f.  3rd s.f.  4th s.f.

⑦ ⑧ ⓪ ②

1st s.f.  2nd s.f.  3rd s.f.

0 . 0 ② ④ ⑤

Placeholders are not significant.

**Worked example**

Round 7802 to 1 significant figure.

<u>7</u> 8 0 2 — Underline the 1st significant figure.

7 ⑧ 0 2 — Look at the next digit. It is more than 5, so round up.

8 0 0 0   (1 s.f.) — Insert 0s to keep the place value.

**Error intervals** are inequalities that show possible values of numbers before rounding.

**Truncate** means removing digits without rounding. 3.7546 truncated to 1 decimal place is 3.7

$x$ is **rounded** to 3.7 (1 d.p.)

The error interval for $x$ is

$3.65 \leq x < 3.75$

3.6   3.65  3.67  3.75  3.8

$n$ is **truncated** to 3.7

The error interval for $n$ is

$3.7 \leq n < 3.8$

3.6   3.65  3.7  3.75  3.8

**LINK**

See inequalities in Chapter 8.

## Error intervals

**Error intervals** use inequalities to show possible values of numbers that have been truncated or rounded.

When a number has been **truncated**, the true value can be up to one unit more than the given measurement.

When a number has been **rounded**, the true value could be half a unit less or up to half a unit more than the stated value.

You can use a **number line** to represent the values that are included in the error interval.

---

### Worked example

**1.** $x$ is **truncated** to 3.7

Draw the error interval for $x$ on a number line.

> Minimum is anything that starts with '3.7'
>
> Maximum must be below 3.8,
> So, error interval is $3.7 \leq x < 3.8$

> Draw your solution on a number line.
>
> A hollow circle shows that the number is <u>not</u> included (less/greater than).
>
> A filled-in circle shows that the number <u>is</u> included (less/greater than or equal to).

**3.** A greetings card has side lengths 11 cm to the nearest centimetre.

Write the error interval showing the maximum and minimum possible lengths of the sides.

$$10.5 \leq x < 11.5$$

> Anything 10.5 cm or above will be rounded up to 11 cm; anything less than 11.5 cm will be rounded down to 11 cm.

**2.** $x$ is **rounded** to 1 decimal place.

The result is 3.7

Draw the error interval for $x$ on a number line.

> Minimum could be half a unit less so $x \leq 3.65$
>
> Maximum can be up to half a unit more, so $x < 3.75$
>
> So, the error interval is $3.65 \leq x < 3.75$

**4.** The mass of an apple, $a$ g, is given as 80 g. Give the error interval for $a$

if the mass was rounded to the nearest 10 g

$$75 \leq a < 85$$

> Mass was rounded to 1 s.f., so error interval will have 2 s.f., and can be greater than or equal to 5 below, and less than 5 above.

---

# ⚙ Knowledge

## 2 Rounding, truncating, error intervals, and estimating

## Error intervals

### Worked example

**1.** $a = \dfrac{F}{m}$.

$F = 45$ to 2 s.f.

$m = 16.1$ to 3 s.f.

By considering bounds, work out the value of $a$ to a suitable degree of accuracy.

$44.5 \leq F < 45.5$ ●——— Find the error intervals for $F$ and $m$.

$16.05 \leq m < 16.15$

lower bound for $a =$ ●——— Find the upper and lower bounds for $a$.

$\dfrac{\text{lower bound of } F}{\text{upper bound of } m} = \dfrac{44.5}{16.15}$

$= 2.755\ldots$

upper bound for $a =$

$\dfrac{\text{upper bound of } F}{\text{lower bound of } m} = \dfrac{44.5}{16.05}$

$= 2.834\ldots$

$a = 2.8$ to 2 s.f. ●——— Both the upper and lower bound of $a$ round to 2.8

> **REVISION TIP**
>
> In the exam, write down the rounded values for each variable, and show each step of your working.

**2.** A racetrack measures <u>800 m</u> to the nearest <u>10 metres</u>. A runner can run around the track in <u>178 s</u> to the <u>nearest second</u>. Calculate the runner's fastest <u>speed</u> to a <u>suitable degree of accuracy</u>. Give a reason for your answer.

Underline the key information and the instruction.

upper bound of distance: 805 m

lower bound of distance: 795 m

upper bound of time: 178.5 s

lower bound of time: 177.5 s

Calculate the upper and lower bounds of the values.

$\text{speed} = \dfrac{\text{distance}}{\text{time}}$

$\text{upper bound of speed} = \dfrac{805}{177.5} = 4.53\ldots \text{ m/s}$

$\text{lower bound of speed} = \dfrac{795}{178.5} = 4.45\ldots \text{ m/s}$

speed $= 4.5$ m/s to 2 s.f., because the upper and lower bounds of the speed are both 4.5 m/s to 2 s.f. ●

To calculate the fastest speed, you want the largest possible solution from the division, so use $\dfrac{\text{upper bound of distance}}{\text{lower bound of time}}$.

## Estimation

When **estimating**, round numbers to 1 significant figure before performing the calculation.

### Worked example

Work out an estimate for the value of $\dfrac{8.762 \times 13.7}{4.82}$

$\dfrac{8.762 \times 13.7}{4.82}$

$\approx \dfrac{9 \times 10}{5}$ ●——— Rewrite the calculation with all numbers rounded to 1 s.f.

$\approx$ means 'approximately equal to'.

$= 18$

## 2 Rounding, truncating, error intervals, and estimating

Learn the answers to the questions below, then cover the answers column with a piece of paper and write as many as you can. Check and repeat.

| | Questions | | Answers |
|---|---|---|---|
| 1 | What is rounding? | *Put paper here* | Writing a number to a given place value. |
| 2 | If the value you are considering for rounding is 5, do you round up or down? | *Put paper here* | Round up. |
| 3 | When you round to a 1 decimal place, which number tells you whether to round up or round down? | *Put paper here* | The number in the second decimal place. |
| 4 | What is the first significant figure of a number? | *Put paper here* | The first non-zero digit, counting from the left. |
| 5 | In the number 56 743, which is the 4th significant figure? | *Put paper here* | 4 (round to the nearest 10). |
| 6 | In the number 0.0 003 654, which is the 2nd significant figure? | *Put paper here* | 6 (round to the nearest 0.00001). |
| 7 | How do you find the error interval if a number has been rounded to a given unit? | *Put paper here* | How do you find the error interval if a number has been rounded to a given unit? |
| 8 | How do you find the error interval if a number has been truncated to a given unit? | *Put paper here* | The minimum is the given measurement, add one unit for the maximum. |
| 9 | When multiplying error intervals, what calculation should you use to find the maximum possible value? | *Put paper here* | upper bound × upper bound = maximum value |
| 10 | What is an error interval? | *Put paper here* | An inequality that shows the range of possible values of unrounded numbers. |

## Previous questions

Now go back and use these questions to check your knowledge of previous topics.

| | Questions | | Answers |
|---|---|---|---|
| 1 | What does commutative mean? | *Put paper here* | Can be done in any order. |
| 2 | What is another word for the result of raising something to the power of −1? | *Put paper here* | Reciprocal. |
| 3 | What is the quotient of two numbers? | *Put paper here* | The result you get when you divide one of the numbers by the other. |
| 4 | What is the value of $(xy)^0$? | *Put paper here* | 1 |

## Exam-style questions

**2.1** When truncated to 1 digit, $P = 1$

When truncated to 1 decimal place, $Q = 0.6$

Work out the error interval for $P + Q$. **[3 marks]**

**2.2** $\sqrt{6} = 2.449$ correct to 4 significant figures.

Work out the value of $\sqrt{0.06}$ correct to 3 significant figures. **[3 marks]**

**2.3**  $A = \dfrac{B}{C^2}$

$B = 6.2$ correct to 2 significant figures

$C = 20.3$ correct to 3 significant figures

**(a)** Calculate the lower bound of $A$ to 5 significant figures. **[3 marks]**

The upper bound for $A$ is 0.015242 to 5 significant figures.

**(b)** By considering bounds, write down the value of $A$ to a suitable degree of accuracy.

Give a reason for your answer. **[2 marks]**

**2.4** A steel block is made in the shape of a cuboid.

The length of the block is 30.0 cm to the nearest mm.

The width of the block is 25.4 cm to the nearest mm.

The height of the block is 9.2 cm to the nearest mm.

The mass of the block, to the nearest 250 g, is 56.5 kg.

By considering bounds, show that the density of the steel is 0.008 kg/cm³, correct to one significant figure.

Show working and give a reason for your final answer. **[5 marks]**

**2.5**  Round 20 193 to

**(a)** 4 significant figures **[1 mark]**

**(b)** 3 significant figures **[1 mark]**

**(c)** 2 significant figures **[1 mark]**

**(d)** 1 significant figure. **[1 mark]**

**2.6** **(a)** Use your calculator to work out $\frac{1}{3}(0.02 \times 11.9)^2$.

Write all the figures on your calculator display. **[1 mark]**

**(b)** Write your answer to part **(a)**

**(i)** truncated to 2 decimal places **[1 mark]**

**(ii)** rounded to 2 significant figures. **[1 mark]**

 **2.7** Akira rounds 0.065 29 to 2 significant figures and gives the answer 0.07.

Akira is wrong. Explain why. **[1 mark]**

 **2.8** Use approximations to work out an estimate for the value of $\frac{2.67 \times 1.36}{0.11 + 0.42}$ **[2 marks]**

 **2.9** A biologist visits a lake at the start of January and works out that the number of fish in the lake is approximately 1000.

She thinks that the population is growing at a rate of 17 fish per day.

Estimate how many fish there will be in the lake five months later. **[3 marks]**

 **2.10** In one week, an Italian restaurant sells 96 portions of lasagne. The restaurant sells a portion of lasagne for £8.95 and each portion costs £3.20 to make.

Estimate the profit the restaurant makes from lasagne in the week. **[3 marks]**

 **2.11** Work out an estimate for the value of each square root. Give your answers to 1 decimal place.

**(a)** $\sqrt{47}$ **[1 mark]**

**(b)** $\sqrt{200}$ **[1 mark]**

**Questions referring to previous content**

**2.12** Write $\frac{\sqrt{5^3}}{5} + 2\sqrt{125}$ in index form, simplified as much as possible. **[4 marks]**

**2.13** Put one or more sets of brackets in the following statement to make it true:

$20 \div 12 - 7 + 8 \times 4 + 1 + 3 = 47$ **[2 marks]**

# ⚙ Knowledge

## 3 Factors, multiples, primes, standard form, and surds

### Factors, multiples, and prime numbers

A **factor** is an **integer** that divides exactly into a number.

3 and 4 are factors of 12.

A multiple is the result obtained when one integer is multiplied by another.

12 is a **multiple** of both 3 and 4.

| factor of 12 | factor of 12 | multiple of 3 and 4 |

For example, multiples of 5 are:

$1 \times 5 = 5$

$2 \times 5 = 10$

$3 \times 5 = 15...$ and so on.

For large numbers we can use **prime factor decomposition** to help us find the HCF and LCM. A **prime number** is a number with exactly two factors. A **prime factor** is a factor that is also a prime number.

The **highest common factor** (**HCF**) of two numbers is the *biggest* number that is a factor of both of the numbers.

The **lowest common multiple** (**LCM**) of two numbers is the *smallest* number that is a multiple of both of the numbers.

> **WATCH OUT**
>
> 1 is not a prime number as it only has one factor (which is 1).

---

**Worked example**

Write 189 as a product of its prime factors.

$189 = 3 \times 3 \times 3 \times 7$

$189 = 3^3 \times 7$

Use a prime factor tree to write a number as a product of its prime factors.

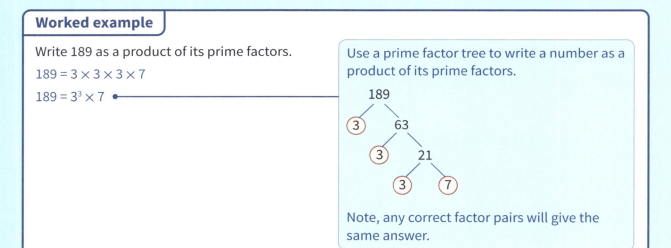

Note, any correct factor pairs will give the same answer.

---

# Factors, multiples, and prime numbers

## Worked example

Find the HCF and LCM of 150 and 225.

$150 = 2 \times 3 \times 5 \times 5$ •⎯⎯⎯⎯⎯⎯⎯⎯⎯

$225 = 3 \times 3 \times 5 \times 5$

> Write 150 and 225 as products of their prime factors.

**REVISION TIP**

Use a factor tree to find the prime factors.

> Draw a Venn diagram of the factors.

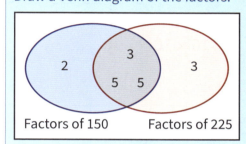

HCF $= 3 \times 5 \times 5 = 75$ •⎯⎯⎯⎯⎯⎯

> **HCF** is the product of numbers in the intersection.

LCM $= 2 \times 3 \times 5 \times 5 \times 3 = 450$ •⎯⎯⎯

> **LCM** is the product of all the numbers.

## Worked example

$x = 2^2 \times 3 \times 5^2$ and $y = 2^3 \times 5 \times 11^2$

(a) Work out the highest common factor of $x$ and $y$.

Give your answer in index form.

HCF $= 2 \times 2 \times 5$ •⎯⎯⎯⎯⎯⎯⎯⎯⎯

$\phantom{HCF} = 2^2 \times 5$

(b) Work out the lowest common multiple of $x$ and $y$.

Give your answer in index form.

LCM $= 5 \times 3 \times 2 \times 2 \times 5 \times 2 \times 11 \times 11$

$\phantom{LCM} = 2^3 \times 3 \times 5^2 \times 11^2$

> Find the prime factors and show them in a Venn diagram.

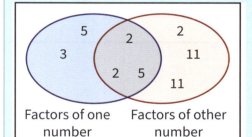

## 3 Factors, multiples, primes, standard form, and surds

### Standard form: The basics

**Standard form** is when a power of 10 is used to rewrite a very large or very small number.

Standard form is written in this way:

| between 1 and 10 i.e. $1 \leq a < 10$ | ●── $(a) \times 10^{(n)}$ ──● | positive or negative integer |

**LINK**
- To remind yourself about multiplying and dividing by powers of 10, look back at Chapter 1.
- To remind yourself about indices, look back at Chapter 2.

These numbers are not written in standard form.

| This number needs to lie in the interval $1 \leq a < 10$ | ●── $16 \times 10^7$ |

$2.4 \times 10^{0.5}$ ●──| This number needs to be an integer |

### Adding or subtracting numbers in standard form

To add or subtract numbers in standard form, convert them to ordinary numbers first.

**Worked example**

Work out $8.9 \times 10^{-4} - 6 \times 10^{-6}$

Give your answer in standard form.

$8.9 \times 10^{-4} = 0.000\,89$

$6 \times 10^{-6} = 0.000\,006$ ●──| Convert to ordinary numbers. |

$$\begin{array}{r} 0.000\,890 \\ -0.000\,006 \\ \hline 0.000\,884 \end{array}$$ ●──| Subtract |

$= 8.84 \times 10^{-4}$ ●──| Convert back to standard form. |

**LINK**
To remind yourself about addition and subtraction, look back at Chapter 1.

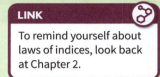 

## Multiplying or dividing numbers in standard form

To multiply or divide numbers in standard form, you need to use these two laws of indices:

1. $a^n \times a^m = a^{n+m}$ — If you multiply numbers with the same base, you add the powers.

2. $a^n \div a^m = a^{n-m}$ — If you divide numbers with the same base, you subtract the powers.

**LINK**

To remind yourself about laws of indices, look back at Chapter 2.

### Worked example

$a = 6 \times 10^5$ and $b = 3 \times 10^{-2}$. Work out the value of:

**1.** $a \times b = (6 \times 10^5) \times (3 \times 10^{-2})$

$= (6 \times 3) \times (10^5 \times 10^{-2})$ ← Group ordinary numbers on one side.

$= 18 \times 10^{5+(-2)}$ ← Use $a^n \times a^m = a^{n+m}$ to simplify the indices.

$= 18 \times 10^3$

$= 1.8 \times 10^4$ ← Give your answer in standard form.

**2.** $a \div b \quad = (6 \times 10^5) \div (3 \times 10^{-2})$

$= \dfrac{6 \times 10^5}{3 \times 10^{-2}}$ ← Write as a fraction.

$= \dfrac{6}{3} \times \dfrac{10^5}{10^{-2}}$ ← Group ordinary numbers and powers of 10.

$= 2 \times 10^{-5-(-2)}$ ← Use $a^n \div a^m = a^{n-m}$

$= 2 \times 10^7$ ← Give your answer in standard form.

In the previous Worked example, part 1, you saw that $18 \times 10^3$ is **not** in standard form.
To make the first number between 1 and 10, you must divide 18 by 10 and multiply $10^3$ by 10.

### Worked example

Write each number in standard form.

**1.** $23 \times 10^4$

$23 \times 10^4$

$\div 10 \quad \times 10$

$2.3 \times 10^5$

← To make the first number between 1 and 10, divide 23 by 10.

← Multiply by 10 to keep the same value.

**2.** $0.023 \times 10^4$

$0.023 \times 10^4$

$\times 100 \quad \div 100$

$2.3 \times 10^2$

← To change 0.023 to 2.3, multiply by 100.

← Divide by 100 to keep the same value.

# ⚙ Knowledge

## 3 Factors, multiples, primes, standard form, and surds

## Surds

A **surd** is an irrational **square root**. Surds can often be simplified using these two laws:

$$\sqrt{ab} = \sqrt{a}\,\sqrt{b} \text{ and } \sqrt{\frac{a}{b}} = \frac{\sqrt{a}}{\sqrt{b}}$$

### Worked example

Simplify fully $\sqrt{45} - \sqrt{32} + \sqrt{5}$

$\sqrt{45} = \sqrt{9}\,\sqrt{5}$

$\quad = 3\sqrt{5}$

> Simplify each surd. Look for a factor that is a square number. ($\sqrt{5}$ can't be simplified)

$\sqrt{32} = \sqrt{16}\,\sqrt{2}$

$\quad = 4\sqrt{2}$

$\sqrt{45} - \sqrt{32} + \sqrt{5}$

> Rewrite the calculation using the simplified surds.

$\quad = 3\sqrt{5} - 4\sqrt{2} + \sqrt{5}$

$\quad = 4\sqrt{5} - 4\sqrt{2}$

> Collect like terms.
>
> This can't be simplified any further.

Rewriting a fraction to remove any surds from the denominator is called rationalising the denominator.

### Worked example

**1.** Simplify $\dfrac{1}{2 + \sqrt{3}}$

$\dfrac{1(2 - \sqrt{3})}{(2 + \sqrt{3})(2 - \sqrt{3})}$

$= \dfrac{2 - \sqrt{3}}{4 - 2\sqrt{3} + 2\sqrt{3} - 3}$

$= \dfrac{2 - \sqrt{3}}{1}$

$= 2 - \sqrt{3}$

> There are **two terms** in the denominator, so to multiply the numerator and denominator by the denominator with the *opposite sign* between the terms.
>
> Expand the brackets.

> These cancel each other out.

> Multiply the numerator and denominator by the denominator with the opposite sign between the terms.

**2.** Write $\dfrac{1 + \sqrt{8}}{1 - \sqrt{2}}$ in the form $a + b\sqrt{2}$, where $a$ and $b$ are integers.

$\dfrac{1 + \sqrt{8}}{1 - \sqrt{2}} = \dfrac{(1 + \sqrt{8})(1 + \sqrt{2})}{(1 - \sqrt{2})(1 + \sqrt{2})}$

$= \dfrac{1 + \sqrt{2} + \sqrt{8} + \sqrt{16}}{1 - 2}$

$= \dfrac{1 + \sqrt{2} + 2\sqrt{2} + 4}{-1}$

$= -5 + 3\sqrt{2}$

> Expand the brackets.

> Simplify any surds.

> Collect like terms and simplify.

> An expression over $-1$ becomes the negative of the expression.

> **Key terms**
>
> Make sure you can write a definition for these key terms.

factor   highest common factor   lowest common multiplier   multiple   prime factor   prime factor decomposition   prime number   root   standard form   surd

## 3 Factors, multiples, primes, standard form, and surds

Learn the answers to the questions below, then cover the answers column with a piece of paper and write as many as you can. Check and repeat.

## Questions | Answers

| | Questions | Answers |
|---|---|---|
| 1 | What is a factor? | An integer that divides exactly into another integer leaving no remainder. |
| 2 | What is a multiple? | A multiple of $a$ is $n \times a$, where $n$ is any other integer. |
| 3 | What does HCF and LCM stand for? | Highest common factor and Lowest common multiple. |
| 4 | What is a prime number? | A number with only two factors, one and itself. |
| 5 | Why is 2 the only even prime number? | All other even numbers are divisible by 2 and therefore have more than two factors. |
| 6 | When a number is written in standard form, what form does it take? | $a \times 10^n$, where $1 \leq a < 10$ and $n$ is an integer. |
| 7 | When a number is written in standard form ($a \times 10^n$) what values can the $a$ take? | $1 \leq a < 10$ |
| 8 | When a number is written in standard form ($a \times 10^n$) what values can the index number, $n$, take? | Any positive or negative integer. |
| 9 | What is standard form used for? | To write very large or very small numbers. |
| 10 | How do you add/subtract numbers that are in standard form? | Convert them to ordinary numbers, add/subtract them and then convert back into standard form. |
| 11 | What is a surd? | An irrational square root. |
| 12 | How do you simplify a surd? | Look for a factor that is a square number. |
| 13 | How do you rationalise the denominator where there is a single surd in the denominator? | Multiply the numerator and denominator by the surd in the denominator. |

*Put paper here*

## Previous questions

Now go back and use these questions to check your knowledge of previous topics.

## Questions | Answers

| | Questions | Answers |
|---|---|---|
| 1 | What is an integer? | A whole number. |
| 2 | What does the letter B in BIDMAS stand for? | Brackets. |
| 3 | What is the first significant figure of a number? | The first non-zero digit, counting from the left. |
| 4 | What does $(a^n)^m$ simplify to? | $a^{nm}$. |

*Put paper here*

# ✏️ Practice

**3.1** Express each of these in the form $a\sqrt{2}$, where $a$ is an integer.

   **(a)** $\sqrt{200} + \sqrt{72} - \sqrt{98}$            **[2 marks]**

   **(b)** $\dfrac{14}{\sqrt{2}}$            **[2 marks]**

**3.2** $\sqrt{3}\,(\sqrt{80} - \sqrt{20})$ can be written in the form $a\sqrt{15}$, where $a$ is an integer. Find the value of $a$.     **[3 marks]**

**3.3** Express each of these in the form $a + b\sqrt{3}$, where $a$ and $b$ are integers.

   **(a)** $\dfrac{8}{2 - \sqrt{3}}$            **[3 marks]**

   **(b)** $\dfrac{\sqrt{3} - 1}{\sqrt{3} + 1}$            **[3 marks]**

**3.4** Show that $(\sqrt{11} - \sqrt{8})\,(\sqrt{11} + \sqrt{8}) = 3$     **[3 marks]**

**3.5** Show that $\dfrac{(\sqrt{2} + \sqrt{50})^2}{3\,(\sqrt{2} - 1)}$ can be written in the form $a\,(b + \sqrt{2})$,

   where $a$ and $b$ are integers.     **[3 marks]**

**3.6** **(a)** Express 750 as a product of its prime factors. Give your answer in index notation.     **[2 marks]**

   **(b)** By considering its prime factors, explain why 750 is not divisible by 4.     **[1 mark]**

**3.7** $x = 2 \times 3^2 \times 7 \times 13$

   **(a)** Is $x$ even or odd? Give a reason for your answer.     **[1 mark]**

   **(b)** Write down the prime factor decomposition for $2x$.     **[1 mark]**

**3.8** $P = 2^3 \times 5 \times 11$         $Q = 2 \times 3^2 \times 5$

   **(a)** Write the highest common factor (HCF) of $P$ and $Q$.     **[1 mark]**

   **(b)** Write the lowest common multiple (LCM) of $P$ and $Q$.     **[1 mark]**

**3.9** $p = 2 \times 3 \times 7 \times x$ $\qquad$ $q = 2^2 \times 5^2$

The highest common factor of $p$ and $q$ is 4.

Work out the value of $x$. **[1 mark]**

**EXAM TIP**

Think about how the HCF links to the prime factor decomposition.

 **3.10** The distance from the Sun to Earth is approximately 150 000 000 km. Write this number in standard form. **[1 mark]**

 **3.11** Put these numbers in order from largest to smallest.

$2.1 \times 10^4$ $\qquad$ $2.3 \times 10^5$ $\qquad$ $0.21 \times 10^4$ $\qquad$ $2200$ **[3 marks]**

**EXAM TIP**

Write all the numbers in the same form.

**3.12** Work out the value of each expression, giving your answers in standard form.

**(a)** $(5 \times 10^4) + (6 \times 10^5)$ **[2 marks]**

**(b)** $(9 \times 10^{-3}) - (3 \times 10^{-4})$ **[2 marks]**

**(c)** $(2.1 \times 10^8) \times (3 \times 10^{-5})$ **[2 marks]**

**(d)** $(8.2 \times 10^3) \div (4.1 \times 10^7)$ **[2 marks]**

 **3.13** A jet can fly at $4 \times 10^3$ km/h.

How long would it take to travel a distance of 3000 km?

Give your answer in minutes. **[3 marks]**

**3.14** A region on a map forms the shape of a rectangle with width $1.2 \times 10^2$ km and length $7 \times 10^3$ km.

Work out the area of this region.

Give your answer in standard form. **[3 marks]**

**3.15** **(a)** Work out $(3.02 \times 10^{18}) \times (6.3 \times 10^{-8})$

Give your answer as an ordinary number. **[2 marks]**

**(b)** Work out $\dfrac{5.3 \times 10^4 - 8.9 \times 10^{-2}}{7.2 \times 10^3}$

Give your answer in standard form. **[2 marks]**

**Questions referring to previous content**

**3.16** Which is greater, and by how much: **[3 marks]**

the sum of $21.263 + 4.801$, rounded to 3 significant figures,

or the difference of $88.155 - 61.379$, truncated to the nearest whole number?

**3.17** Estimate the value of the following expression, by first rounding each number to 1 significant figure:

$\dfrac{4.2^2 \times 5.075}{9.88} + \dfrac{11.7}{2.66}$ **[2 marks]**

# ⚙ Knowledge

## 4 Fractions, decimals, percentages

### Fractions

The **numerator** is the top number in a **fraction**.

The **denominator** is the bottom number in a fraction.

$\dfrac{3}{5}$ — numerator / denominator

An **improper fraction** has a numerator larger than the denominator. For example, $\dfrac{15}{7}$

A **mixed number** has a whole part and a fraction part. For example, $2\dfrac{1}{3}$

---

**Worked example**

**1.** Write $4\dfrac{3}{7}$ as an improper fraction.

$4 = \dfrac{4 \times 7}{7}$ — Convert the whole part into an improper fraction by multiplying by the denominator.

$= \dfrac{28}{7}$

$= \dfrac{28}{7} + \dfrac{3}{7}$ — Add the answer for the whole part to the fraction part.

$= \dfrac{31}{7}$

**2.** Write $\dfrac{17}{6}$ as a mixed number.

$\dfrac{17}{6} = 2$ remainder $5$ — Divide the numerator by the denominator.

$= 2\dfrac{5}{6}$ — Write this as a whole number and a fraction.

---

To compare fractions with different denominators, convert the fractions to **equivalent fractions** that have a common denominator or common numerator.

---

**Worked example**

Put these fractions in ascending order: $\dfrac{2}{3}, \dfrac{5}{12}, \dfrac{5}{8}$

$3 = 1 \times 3$

$12 = 3 \times 2 \times 2$ — Find the LCM of 3, 12, and 8.

$8 = 2 \times 2 \times 2$

$\text{LCM} = 2 \times 2 \times 2 \times 3$

$= 24$

**LINK** 🔗
To remind yourself about the HCF and LCM, look back at Chapter 3.

$\dfrac{2}{3} = \dfrac{16}{24}$

$\dfrac{5}{12} = \dfrac{10}{24}$ — Convert the fractions to 24ths by identifying the number the denominator needs to be multiplied by to get to 24, and multiplying the numerator by the same number.

$\dfrac{5}{8} = \dfrac{15}{24}$

$\dfrac{10}{24}, \dfrac{15}{24}, \dfrac{16}{24}$ — Put the fractions in order by comparing the numerators (smallest first).

$\dfrac{5}{12}, \dfrac{5}{8}, \dfrac{2}{3}$ — Write the fractions in their original form in the order corresponding to the previous step.

# Calculations with fractions

## Multiplying fractions

Multiply the numerators and multiply the denominators.

- $\frac{4}{5} \times \frac{2}{3} = \frac{2 \times 4}{5 \times 3} = \frac{8}{15}$

Work out these calculations. Give your answers in their simplest form.

**(a)** $\frac{3}{4}$ of $\frac{5}{6}$

$\frac{3}{4}$ of $\frac{5}{6} = \frac{3}{4} \times \frac{3}{10}$ •———— Simplify the fraction.

$= \frac{9}{40}$

$= \frac{5}{8}$

**(b)** $2\frac{7}{9} \times \frac{3}{10}$

$2\frac{7}{9} \times \frac{3}{10} = \frac{25}{9} \times \frac{3}{10}$ •———— Convert $2\frac{7}{9}$ to an improper fraction.

$= \frac{75}{90}$

$= \frac{5}{6}$

**REVISION TIP**

After all fraction calculations, use equivalent fractions to simplify your answer if possible.

## Dividing fractions

To divide by a fraction, multiply by its reciprocal.

$\frac{4}{5} \div \frac{2}{3} = \frac{4}{5} \times \frac{3}{2} = \frac{4 \times 3}{5 \times 2} = \frac{12}{10}$

**Worked example**

Work out these calculations. Give your answers in their simplest form.

**(a)** $\frac{3}{8} \div \frac{7}{10}$

$\frac{3}{8} \div \frac{7}{10} = \frac{3}{8} \times \frac{10}{7}$ •——— Swap to a multiplication by changing the sign and swapping the numerator and denominator of the second fraction.

$= \frac{30}{56}$

$= \frac{15}{28}$

Multiply the numerators and the denominators, and simplify.

**(b)** $5 \div \frac{1}{3}$

$5 \div \frac{1}{3} = 5 \times 3$

$= 15$

**REVISION TIP**

The **reciprocal** of a number is the the number you would have to multiply it by to get the answer 1. For example, the reciprocal of 2 is $\frac{1}{2}$.

## Adding and subtracting fractions

Change any mixed numbers into improper fractions. Write both fractions over a common denominator. Add or subtract the numerators.

$\frac{4}{5} + \frac{2}{3} = \frac{12}{15} + \frac{10}{15} = \frac{22}{15}$

**Worked example**

**1.** Two pipes of lengths $2\frac{3}{4}$ m and $1\frac{2}{3}$ m are joined end to end. How long is the new length of pipe?

$2\frac{3}{4}m + 1\frac{2}{3}m = \frac{11}{4}m + \frac{5}{3}m$ •——— Rewrite mixed numbers as improper fractions.

$= \frac{33}{12}m + \frac{20}{12}m$ •——— Find the LCM of the denominators (12) and write the fractions with this as the new denominator.

$= \frac{53}{12}m$

$= 4\frac{5}{12}m$ •——— Add fractions and convert answer to a mixed number.

**WATCH OUT**

It's usually easiest to write mixed numbers as improper fractions before you add or subtract them.

# ⚙ Knowledge

## 4 Fractions, decimals, percentages

## Converting between fractions, decimals, and percentages

**Percent** means "out of 100".

**Percentages** can be written as fractions with 100 on the denominator.

Percentages can be converted into decimals or fractions, and vice versa.

$$40\% = \frac{40}{100} \text{ or } \frac{2}{5} = 0.4$$

---

**Worked example**

Niamh has some money. She spends $\frac{2}{5}$ on shoes and 35% on clothes.

What fraction of the original amount does she have left?

$$35\% = \frac{35}{100}$$ — Convert 35% to a fraction.

$$\frac{2}{5} = \frac{40}{100}$$ — Write $\frac{2}{5}$ as an equivalent

$$\frac{35}{100} + \frac{40}{100} = \frac{75}{100}$$ — fraction with denominator 100.

has been spent.

$$1 - \frac{75}{100} = \frac{100}{100} - \frac{75}{100}$$ — Add the fractions to find the total amount spent.

$$= \frac{25}{100}$$ — Subtract the fraction spent from 1 to find the fraction that is left.

$$= \frac{1}{4}$$ — Simplify the fraction.

---

**Worked example**

There are 16 balls in a bag and 5 of them are red. What percentage of the balls are red?

$\frac{5}{16}$ of the balls in the bag are red. — Write the fraction of red balls in the bag.

Convert this to a decimal by dividing.

$$0.3125 \times 100 = 31.25$$ — Multiply by 100 to convert to a percentage.

31.25% of balls are red.

---

A recurring decimal is a decimal number which has a repeating number or pattern of numbers after the decimal point. A dot over the recurring numbers show they are recurring.

$$5.333333 = 5.\dot{3} \qquad\qquad 3.717171 = 3.\dot{7}\dot{1}$$

Recurring decimals can be converted into fractions.

# Converting between fractions, decimals, and percentages

## Worked example

Write $0.\dot{7}\dot{2}$ as a fraction in its simplest form.

Let $x = 0.7272...$

$100x = 72.7272...$

> Two digits recur, so multiply by 100.

$100x - x = 72.727\,272 - 0.7272$

$99x = 72$

$x = \dfrac{72}{99}$

$x = \dfrac{8}{11}$

> Clear the decimal places by subtracting $x$ from $100x$.

> Divide both sides by 99 to find $x$.

### REVISION TIP

If only the first digit was recurring, you'd multiply by 10 and subtract to give $9x$.

## Worked example

Prove algebraically that $0.5\dot{8}$ converts to the fraction $\dfrac{53}{90}$

### WATCH OUT

This example is slightly different because the first digit after the decimal point (5) doesn't recur.

Let $x = 0.5888...$

$10x = 5.888...$

$100x = 58.888...$

> Start by multiplying $0.5888..$ by 10 so only the recurring digits appear after the decimal point.

$100x - 10x$
$= 58.888... - 5.888...$

$90x = 53$

$x = \dfrac{53}{90}$

> Multiply by 100, and subtract $90x$ to give $10x$, then solve.

## Percentages

Percentages of amounts can be calculated using **multipliers**.

original amount × multiplier = new amount

### Worked example

Work out 63% of 576.

$63\% = \dfrac{63}{100}$

> Find the decimal equivalent of 63%

$= 0.63$

$= 576 \times 0.63$

> Multiply the amount by the multiplier (you can use a calculator for this).

$= 362.88$

Amounts can be increased or decreased by a given percentage.

### Worked example

**1.** Increase 75 by 8%

$100\% + 8\% = 108\%$

> Add the percentage increase to 100%

$108\% = \dfrac{108}{100}$

$= 1.08$

> Find the multiplier.

$75 \times 1.08 = 81$

> Multiply the original amount by the multiplier.

**2.** Decrease 75 by 8%

$100\% - 8\% = 92\%$

> Subtract the percentage from 100%

$92\% = \dfrac{92}{100}$

$= 0.92$

> Find the multiplier.

$75 \times 0.92 = 69$

> Multiply the original amount by the multiplier.

# Knowledge

## 4 Fractions, decimals, percentages

### Percentages

**Worked example**

The price of an armchair is £52.34 including VAT.

Work out the original price of the armchair before VAT.

$$52.34 = 120\% \text{ of } x$$

$$52.34 = 1.2x$$

$$x = \frac{52.34}{1.2}$$

$$x = 43.61666\ldots$$

original price = £43.62

> Let £$x$ be the original price. Find an expression for $x$.

> Rewrite the expression using a multiplier ($120\% = 1.2$) and solve for $x$.

> **REVISION TIP**
> VAT is 20% of the original price. It is added on to the cost of the item.

You might be given an amount before and after a change and asked to calculate the percentage increase or decrease.

$$\text{percentage increase} = \left(\frac{\text{actual increase}}{\text{original amount}}\right) \times 100\% \qquad \text{percentage decrease} = \left(\frac{\text{actual decrease}}{\text{original amount}}\right) \times 100\%$$

**Worked example**

The height of a plant increases from 64 cm to 87.04 cm.

Work out the percentage increase in the height of the plant.

**Method 1: Using the formula**

$$87.04 - 64 = 23.04 \text{ cm}$$

$$\text{percentage increase} = \frac{23.04}{64} \times 100\%$$

$$= 0.36 \times 100\%$$

$$= 36\%$$

> Subtract the starting amount from the end amount.

> Divide the difference by the starting amount and multiply by 100 to find the percentage increase.

**Method 2: Using multipliers**

$$64x = 87.04$$

$$x = \frac{87.04}{64}$$

$$x = 1.36\text{, so the percentage increase in height is 36\%}$$

> Write an equation that includes the start and end amounts with $x$ as the multiplier.

> Rearrange to find the value of $x$.

> **REVISION TIP**
> A 1 before the decimal place = an increase. (original amount plus a bit extra)
>
> A 0 before the decimal place = a decrease (less than the original amount)

## Percentages

To increase or decrease an amount, A, by the same percentage $n$ times, use $A = \text{multiplier}^n$

This is used to calculate **compound interest**. This is where the interest is left in the account, so interest is earned on the original amount plus the interest.

The other type of interest is **simple interest**. This is where you only earn interest on the initial investment, or the interest is taken out at the end of each year.

> **Worked example**
>
> Natalie invests £3500 in a savings account for five years at an interest rate of 4% per year. She needs to decide whether to take the interest out of the account each year or leave it in. Work out how much extra interest she will earn if she leaves it in for the full five years.
>
> 4% of £3500 = 0.04 × £3500
> = £140
>
> £140 × 5 = £700
>
> total after 5 years = $3500 \times 1.04^5$
> = £4258.29
>
> £4258.29 − £3500 = £758.29
>
> £758.29 − £700 = £58.29
>
> If the interest is taken out at the end of each year, it is **simple interest**. Calculate the interest earned in one each year.
>
> Multiply by 5 to calculate the amount of interest earned over five years.
>
> If the interest is left in, it is **compound interest**.
> Convert the interest percentage to a multiplier (4%, so 1.04)
> Use the number of years as the power.
> Substitute into the compound interest equation to find the total in the account after five years.
>
> To find the interest earned, subtract the initial investment.
>
> Subtract the simple interest amount from the compound interest amount to find the difference.

**Key terms** Make sure you can write a definition for these key terms.

compound interest   denominator   equivalent fraction   fraction   improper fraction   multiplier   numerator   percent   percentage   reciprocal   simple interest

## 4 Fractions, decimals, percentages

Learn the answers to the questions below, then cover the answers column with a piece of paper and write as many as you can. Check and repeat.

| Questions | Answers |
|---|---|
| **1** What is an improper fraction? | A fraction where the numerator is larger than the denominator. |
| **2** What is a mixed number? | A number that has a whole number and a fraction. |
| **3** What is the reciprocal of a number? | The reciprocal of a number is 1 over the number. It is the value that you multiply a number by to get 1. |
| **4** How do you multiply two fractions? | Multiply the numerators together and multiply the denominators together. |
| **5** How do you divide by a fraction? | Multiply by its reciprocal. |
| **6** How do you add or subtract fractions? | Write all the fractions with common denominators and add or subtract the numerators. |
| **7** How do you add or subtract mixed numbers? | Change into improper fractions, then write all the fractions with common denominators, then add the numerators. |
| **8** What form should you leave a fractional answer in? | As a fully simplified proper fraction or mixed number. |
| **9** How can you find the fraction of an amount? | Multiply the fraction by the amount. |
| **10** How do you convert a percentage into a decimal? | Divide the percentage by 100. |
| **11** How do you change a percentage into a fraction? | Write the percentage as a fraction with 100 as the denominator. |
| **12** What is simple interest? | A fixed amount of interest calculated as a percentage of the original amount. |
| **13** What is compound interest? | An amount of interest that is calculated as a percentage of the current amount. |

*(Put paper here)*

### Previous questions

Now go back and use these questions to check your knowledge of previous topics.

| Questions | Answers |
|---|---|
| **1** What is rounding? | Writing a number to a given place value. |
| **2** Describe how you would simplify $a^5 \div a^4$? | Subtract the powers. |
| **3** What does BIDMAS tell you? | The order in which to do a calculation. |

*(Put paper here)*

# Practice

**4**

## Exam-style questions

 **4.1** Put these numbers in descending order:

16%      $\frac{1}{6}$      0.165      $\frac{17}{100}$      **[2 marks]**

 **4.2** May buys a caravan.

The caravan costs £26 759 plus VAT at 20%.

May pays a deposit.

She pays the remainder in 18 equal payments of £1450.60

Work out how much May pays for the deposit.      **[4 marks]**

 **4.3** Maxwell is reading a book on his e-reader.

When he picks it up one day, it tells him he is $\frac{1}{3}$ of the way through the book.

He reads until he is $\frac{3}{4}$ of the way through.

What fraction of the book does he read?

**[2 marks]**

 **4.4** Work out the perimeter of this triangle.

 **4.5** A triangle has base $1\frac{1}{5}$ cm and height $\frac{6}{5}$ cm.

A rectangle has width $\frac{2}{5}$ cm and length $x$ cm.

The rectangle and the triangle have the same area.

Work out the value of $x$.

**[3 marks]**

**4.6** Vasiliki has a piece of material $3\frac{3}{4}$ m long.

She cuts it into smaller pieces of length $\frac{5}{6}$ m.

How many smaller pieces can she make.

What fraction of a metre will be left over?

**Number of smaller pieces = _____**

**Fraction left = _____** **[2 marks]**

> **EXAM TIP**
> This question combines fractions with algebra and geometry. Find the area of the triangle, then write and solve an equation.

 **4.7** Autumn ran some races.

15% of her races were 5 km runs, $\frac{7}{10}$ of her races were 10 km runs and the rest were half marathons.

She ran 20 races in total.

How many of her races were half marathons? **[3 marks]**

**4.8** Express these fractions as decimals.

**(a)** $\frac{1}{18}$ **(b)** $\frac{20}{33}$ **[2 marks]**

 **4.9** Prove algebraically that $0.\dot{5}$ converts to the fraction $\frac{5}{9}$ **[2 marks]**

> **EXAM TIP**
> Let $x = 0.\dot{5}$ .and find $10x$

 **4.10** Convert $0.6\dot{4}$ to a fraction in its simplest form. **[2 marks]**

 **4.11** Prove algebraically that $0.0\dot{5}\dot{6} = \frac{28}{495}$ **[3 marks]**

**4.12** A baby lizard is 6 cm long.

When it is adult, the lizard is 22 cm long.

Work out the percentage increase in the length of the lizard.

Give your answer to 3 significant figures. **[3 marks]**

**4.13** Darren runs a marathon in 3 hours and 15 minutes.

After some training, he runs it in 3 hours.

Calculate the percentage decrease in the time it takes Darren to run a marathon. **[2 marks]**

> **EXAM TIP**
> Convert the times into minutes first.

 **4.14** Farida pays £150 for a box of 10 000 screws.

She packs them into smaller boxes containing 1250 screws per box.

She sells all of these boxes for £24 per box.

Work out Farida's percentage profit. **[4 marks]**

**4.15** The price of a car was increased by 50% and then reduced by 25%.

Work out the overall percentage increase on the original price. **[3 marks]**

 **4.16** The numerator of a fraction is increased by 48%.

Its denominator is decreased by $87\frac{1}{2}$%.

The resulting fraction is $\frac{37}{42}$

Work out the original fraction. Give the fraction in its simplest form. **[3 marks]**

 **4.17** Two banks have an interest rate of 6% per annum.

Bank A pays compound interest and Bank B pays simple interest.

**(a)** In January 2023, Keysha invests £2450 in Bank A and Paul invests £2450 in Bank B.

How much more money will Keysha have than Paul in January 2030? **[4 marks]**

**(b)** Phoebe invests money in Bank A.

She needs at least £5000 in her account by January 2033.

Work out how much, in whole pounds, Phoebe needs to invest in January 2023 to reach this figure. **[3 marks]**

 **4.18** Avani invests £6000 in an account for one year.

At the end of the year, interest is added to her account.

Avani pays tax on the interest at a rate of 15%. She pays £19.80 tax.

Work out the percentage interest rate for the account. **[3 marks]**

 **4.19** The value of an oil painting increases at a rate of $x$% per year.

Initially, the painting is worth £4000.

After five years, it is worth £4300

Work out the value of $x$ to 1 decimal place. **[3 marks]**

 **4.20** Ehmet bought a flat. Its value fell by 3% in the first year.

During the next two years, the value went up by 7% per year.

At the end of the third year, his flat was worth £285 000

**(a)** Calculate the percentage change in the value of the flat to 3 significant figures. **[3 marks]**

**(b)** Work out how much Ehmet paid for his flat. **[2 marks]**

**Questions referring to previous content**

 **4.21** By drawing a Venn diagram, or by any other method, find the HCF and LCM of 300 and 840. **[4 marks]**

 **4.22** Show that $\left(\sqrt{3} + \sqrt{27}\right)^2 = 147$ **[4 marks]**

## 5 Algebra basics

## Algebraic expressions

**Variable** A letter in an expression or equation, such as $x$ or $y$

**Term** A number or letter by itself, or multiple numbers and letters multiplied together, for example, $3x, -\frac{y}{2}, 2ab^2, -3$

**Expression** A collection of letters and numbers with no equals sign, such as $2x + 3y - 4$

**Equation** Two expressions that are connected by an equals sign, such as $2x + 4 = 3x + 1$

Variable Variable
$2\overset{\frown}{x}$ + 4 = $3\overset{\frown}{x}$ + 1
Term Term Term Term
Expression Expression
Equation

## Simplifying expressions

To **simplify** an expression, collect together like terms then add or subtract them according to their signs.

**Worked example**

1. Simplify $3x - 6y + 3x + 2y + 7$

$(3x)(-6y)(+3x)(+2y)(+7)$

$= (3x + 3x)(-6y + 2y)(+7)$

$= (6x)(+4y)(+7)$

→ Collect like terms.

→ Simplify.

2. Simplify $2a^3 b - 3b^2 + a^3 b + 3b^2 + 5ab + 3a^2$

$(2a^3b)(-3b^2) + (a^3b)(+3b^2)(+5ab)(+3a^2)$ — Collect like terms.

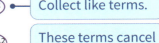

$= (2a^3b + a^3b)(-3b^2 + 3b^2)(+5ab)(+3a^2)$ — These terms cancel each other out.

$= (3a^3b)(+5ab)(+3a^2)$ — Simplify.

You can also simplify expressions that include powers or indices.

**Worked example**

1. Simplify

$\dfrac{12s^5 t^2}{24st^7}$

$\dfrac{12}{24} = \dfrac{1}{2}$ — Simplify the coefficients.

$\dfrac{s^5}{s} = s^4$ — Simplify the $s$ terms.

$\dfrac{t^2}{t^7}$ or $\dfrac{1}{t^5} = t^{-5}$ — Simplify the $t$ terms.

$\dfrac{12s^5 t^2}{24st^7}$ — Put the terms together.

$= \dfrac{s^4}{2t^5}$ or $\dfrac{1}{2}s^4 t^{-5}$

2. Simplify:

$(3rs^2)^3$

$3^3 = 27$ — Cube each of the terms.
$r^3$
$(s^2)^3 = s^{2 \times 3} = s^6$

$27r^3 s^6$ — Put the terms together.

**REVISION TIP**

A coefficient is the number before a letter, e.g. the coefficient of $6m$ is 6.

# Expanding and factorising single brackets

To **expand** a single bracket, multiply each term inside the bracket by the term in front of the bracket.

> **Worked example**
>
> Expand $3x(x + 7)$
>
> **Method 1: Using an area model**
>
> | × | $x$ | $+ 7$ |
> |---|---|---|
> | $3x$ | $3x^2$ | $+ 21x$ |
>
> $= 3x^2 + 21x$
>
> Use the grid to multiply each term inside the bracket by the term outside the bracket.
>
> **Method 2: Drawing lines**
>
> $3x(x + 7)$
>
> $= 3x^2 + 21x$
>
> Use lines to help you multiply each term inside the bracket by the term outside the bracket.

**Factorising** is the opposite of expanding.

To factorise an expression:

- find the highest common factor (HCF) of the terms and put this outside the bracket

- work out what needs to go in the brackets.

> **Worked example**
>
> Factorise fully $12x^2y + 16xy^2$
>
> HCF of 12 and 16 is 4.
>
> HCF of $x^2y$ and $xy^2$ is $xy$
>
> so HCF $= 4xy$.
>
> Find the HCF of the two terms. Put this outside the brackets.
>
> $4xy(\underline{\quad} + \underline{\quad})$
>
> $\frac{12x^2y}{4xy} = 3x$ and $\frac{16xy^2}{4xy} = 4y$
>
> Divide each of the terms by the HCF.
>
> $12x^2y + 16xy^2 = 4xy(3x + 4y)$
>
> Add these terms inside the brackets.

# Substitution

**Substitution** means replacing a variable with a number.

> **Worked example**
>
> Work out the value of $abc + a^2$ when $a = -4$, $b = 5$ and $c = 2$.
>
> $a \times b \times c + a^2$
>
> $= (-4) \times (5) \times (2) + (-4)^2$
>
> Substitute all values.
>
> $= (-4) \times (5) \times (2) + 16$
>
> $= -40 + 16$
>
> $= -24$
>
> Carry out the calculation in BIDMAS order.

**REVISION TIP**

Remember that squaring a negative gives a positive.

# ⚙ Knowledge

## 5 Algebra basics

## Solving linear equations

To **solve** an equation, use **inverse operations** to move the unknown to one side on its own. Inverse operations are operations that are the opposite of each other. Examples include:

| + and − | × and ÷ | $x^2$ and $\sqrt{x}$ |

Keep the equation balanced by doing the same operation to both sides.

> **REVISION TIP** ☑
>
> In a linear equation, the highest power of the variable is always 1. $x + 5 = 7x$ is linear, but $x^2 + 2 = 3x$ is not.

### Worked example

1. Solve $\dfrac{3x}{2} + 4 = 12$

$\dfrac{3x}{2} = 8$ — Subtract 4 from both sides.

$3x = 16$ — Multiply by 2.

$x = \dfrac{16}{3}$ — Divide by 3.

2. Solve $\dfrac{2(x + 3)}{3} = x - 1$

$\dfrac{\cancel{3} \times 2\,(x + 3)}{\cancel{3}} = 3\,(x - 1)$ — Get rid of the denominator by multiplying.

$2x + 6 = 3x - 3$ — Expand the brackets.

$2x + 6 - 2x = 3x - 3 - 2x$

$6 = x - 3$ — Use inverse operations to get $x$ on only one side of the equation.

$6 + 3 = x - 3 + 3$

$x = 9$ — Use inverse operations to get $x$ on its own.

3. $a^3 \times = \dfrac{a^{2n}}{a^4} = a^{-5}$

Find the value of $n$.

$a^{-5} = a^3 \times \dfrac{a^{2n}}{a^4}$ — Use laws of indices to simplify RHS.

$= a^3 \times a^{2n} \times a^{-4}$

$= a^{3 + 2n - 4}$

$= a^{2n - 1}$ — Equate the powers and solve for $n$.

$-5 = 2n - 1$

$2n = -4$

$n = -2$

> **REVISION TIP** ☑
>
> For this question, you need to use laws of indices to help you form a linear equation in terms of $n$.

For example, when solving $2x + 5 = 8 + x$:

A **formula** shows how one variable relates to another variable. For example:

- area = length × width
- $y = 2x + 5$
- distance = speed × time

Inverse operations can be used to rearrange formulae.

### Worked example

1. The formula for the area of a circle is $A = \pi r^2$

   Rearrange the formula to to make $r$ the subject.

   $\dfrac{A}{\pi} = r^2$ — Divide both sides by $\pi$.

   $r = \sqrt{\dfrac{A}{\pi}}$ — Take the square root of both sides.

2. The area of a circle is 2.27 cm².

   Work out the radius of this circle, correct to 1 d.p.

   $r = \sqrt{\dfrac{28.27}{\pi}}$ — Substitute the area into the formula and solve to find $r$.

# Solving linear equations

## Worked example

The square and the triangle have the same perimeter.

13 cm    $(x+5)$ cm

$x$ cm    $x$ cm

Work out the value of $x$.

> **REVISION TIP**
>
> If a question asks you to find the value of $x$ but doesn't state an equation, use the information given to first form, and then solve, an equation.

perimeter of square $= 4x$

perimeter of triangle $= x - (x+5) + 13$ — Find expressions for the perimeters.

$= 2x + 18$ — Set the two expressions equal to each other.

$4x = 2x + 18$

$4x - 2x = 2x - 2x + 18$ — Simplify and solve for $x$.

$2x = 18$

$\dfrac{2x}{2} = \dfrac{18}{2}$ — Subtract $2x$.

$x = 9$ — Divide by 2.

An **identity** is an equation that is always true, no matter what the value of the variable is.

$2x \equiv x + x$

$2a \times b \equiv a \times 2b$

Sometimes, you will need to prove that an identity holds for any value of the variable.

The 'three-line' equals sign ($\equiv$) tells you that this is an identity. It is true no matter what the value of $a$.

You need to show that you can rearrange the left-hand side to make it the same as the right-hand side.

## Worked example

1. Rearrange $\dfrac{1}{a} = \dfrac{1}{b} + \dfrac{1}{c}$ to make $b$ the subject.

$\dfrac{\cancel{a}bc}{\cancel{a}} = \dfrac{a\cancel{b}c}{\cancel{b}} + \dfrac{ab\cancel{c}}{\cancel{c}}$ — Multiply every term by the LCM of the denominators ($abc$) and simplify.

$bc = ac + ab$

$bc - ab = ac$ — Subtract $ab$ so all terms involving $b$ are on the same side of the equation.

$b(c - a) = ac$ — Factorise.

$b = \dfrac{ac}{c-a}$ — Divide by $c - a$

2. Show that $3(a + 2b) - 4a \equiv -a + 6b$

LHS $\equiv 3(a + 2b) - 4a$ — Expand the bracket.

LHS $\equiv 3a + 6b - 4a$

$\equiv -a + 6b$ — Collect like terms.

$\equiv$ RHS

**Key terms**

Make sure you can write a definition for these key terms.

equation   expand   expression   factorise   formula
identity   inequality   inverse operation   like terms
simplify   solve   substitute   term   variable

## 5 Algebra basics

Learn the answers to the questions below, then cover the answers column with a piece of paper and write as many as you can. Check and repeat.

| Questions | Answers |
|---|---|
| **1** In algebra what is a term? | A single number or variable, or numbers and variables multiplied together e.g. 2, $x$, $5y$ are all examples of terms. |
| **2** What are terms of the same type called? | Like terms. |
| **3** What is the opposite of factorising? | Expanding. |
| **4** What does it mean to expand a single bracket? | To multiply each term inside the bracket by the term in front of the bracket. |
| **5** What is an expression? | A collection of letters and numbers which cannot be solved (no equals or inequality sign). |
| **6** What is a coefficient? | The number before a letter, e.g. the coefficient of $6m$ is 6. |
| **7** What is a variable? | A letter used to represent an unknown numerical value. |
| **8** What does $2a$ mean? | 2 multiplied by $a$. |
| **9** What does $a^n$ mean? | $a$ raised to the power of $n$. |
| **10** What does it mean to factorise fully? | To make sure that you have used the highest common factor, not just a common factor. |
| **11** What is the difference between an equation and an identity? | An identity is true for all values of the variable. |
| **12** What is a 'linear' equation? | An equation with no power higher than 1. |

*Put paper here*

## Previous questions

Now go back and use these questions to check your knowledge of previous topics.

| Questions | Answers |
|---|---|
| **1** How do you convert a decimal into a percentage? | Multiply by 100. |
| **2** What is the reciprocal of $x$? | $\frac{1}{x}$ |
| **3** What does HCF stand for? | Highest common factor. |

*Put paper here*

# Practice

## Exam-style questions

**5.1** Simplify

    **(a)** $(2a)^3$ **[2 marks]**

    **(b)** $\dfrac{6x^2y^{-3}}{18yx^{-1}}$ **[2 marks]**

    **(c)** $\sqrt{x^4y^6}$ **[2 marks]**

> **EXAM TIP**
> Convert the square root to index form.

 **5.2** Expand and simplify
$5(3 - x) - 4(6 - 3x)$ **[2 marks]**

 **5.3** Solve

    **(a)** $\dfrac{5 - x}{2} = 12$ **[2 marks]**

    **(b)** $\dfrac{2}{y} = 5$ **[2 marks]**

    **(c)** $3 + p = 4p - 6$ **[2 marks]**

    **(d)** $3(3 - 2p) = 4 - 11p$ **[2 marks]**

 **5.4** Solve
$5y - 2 = \dfrac{y - 13}{2}$ **[3 marks]**

 **5.5** Work out the value of $x$.

    **(a)** $\dfrac{3^{-2} \times 3^8}{3^7} = 3x$ **[3 marks]**

    **(b)** $2^5 \times 4^2 = 8x$ **[3 marks]**

 **5.6** Find the value of $t$, where $\dfrac{c^t \times c^{2t}}{c} = c^2$ **[4 marks]**

 **5.7** Rosalind thinks of a number.

She multiplies her number by two, adds three and then writes the answer.

Next, she multiplies her original number by three, subtracts four and writes the answer.

Rosalind realises she has written the same answer both times.

What is Rosalind's number?

You **must** show your working. **[3 marks]**

> **EXAM TIP**
> Write and solve an equation using $x$ as your unknown number.

# Exam-style questions

**5.8** Show that $(4x - 1)(2x + 5)(3x - 2)$ can be written in the form
$ax^3 + bx^2 + cx + d$ where $a$, $b$, $c$ and $d$ are integers. **[2 marks]**

**5.9** Factorise fully.

**EXAM TIP**

When you're told to factorise 'fully', make sure you take out the HCF.

    **(a)**    $16x + 12xy$ **[2 marks]**

    **(b)**    $x^2y + y^2x$ **[2 marks]**

    **(c)**    $8p - 4p^2q + 6pq$ **[2 marks]**

**5.10** Make $x$ the subject of each formula.

    **(a)**    $2x^2 - 3 = y$ **[2 marks]**

    **(b)**    $a\sqrt{x} + b = c$ **[2 marks]**

**5.11** The power, $P$ watts, in a circuit is given by the formula

$$P = I^2R$$

where $I$ is the current in amps and $R$ is the resistance in ohms.

Rearrange the formula to make $I$ the subject. **[2 marks]**

**5.12** Min-Jun is rearranging a formula to make $x$ the subject. His steps are shown.

Give reasons why Min-Jun's method is wrong.

$$y = a^2 x - b$$
$$y + b = a^2 x$$
$$\sqrt{y + b} = ax$$
$$\frac{\sqrt{y + b}}{a} = x$$

**[3 marks]**

**5.13** Make $r$ the subject of the formula. $6d = \dfrac{n - 7}{3r + 1}$ **[3 marks]**

**5.14** Make $p$ the subject of each formula.

    **(a)**    $mp - q = ap$ **[3 marks]**

    **(b)**    $\dfrac{1}{p} + \dfrac{1}{r} = \dfrac{1}{t}$ **[3 marks]**

**Questions referring to previous content**

 **5.15** An online bank account has a six-character access code.

The first character must be a letter (A to Z).

The remaining characters must be single digit (0 to 9).

Numbers may be used more than once.

**(a)** Work out how many possible access codes there are. **[2 marks]**

The rules for access codes are changed.

Under the new rules, both the first and last characters of the code must be letters. The remaining characters must be single digit numbers. No number or letter may be used more than once. **[2 marks]**

**(b)** How many possible six-character access codes are there now? **[2 marks]**

 **5.16** Write 0.78$\dot{3}$ as a fraction. **[3 marks]**

# Knowledge

## 6 Linear graphs

## Graph terminology

A **coordinate grid** is a grid and a set of **axes** that you can plot points (called **coordinates**) and lines on.

The $x$-axis is a horizontal scale on the coordinate grid. The $x$-**intercept** is a point where a **graph** cuts the $x$-axis. In this graph, it is at $x = 3$.

The $y$-axis is the vertical scale. The $y$-intercept is a point where a graph cuts the $y$-axis. In this graph, it is at $y = 2$.

The **origin** is the point with coordinates (0, 0).

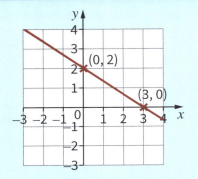

## Linear graphs

A **linear** graph is a set of coordinates that make a straight line when plotted. This line can be horizontal, vertical or sloping.

$y = 3$ is a horizontal line

$y = 3$
goes through
the $y$-axis at (0, 3)

All lines of the form $y = a$ are horizontal for any value of $a$.

$x = 3$ is a vertical line

$x = 3$
goes through
the $x$-axis at (3, 0)

All lines of the form $x = a$ are vertical for any value of $a$.

## Gradients

The **gradient** is a measure used to describe how steep a line graph is. It is how many units up (or down) the line goes for every one unit across.

To find the gradient of a line, first find two points on the line where you can read the exact coordinates. Then:

$$\text{gradient} = \frac{\text{change in } y}{\text{change in } x}$$

Which, in this example $= \frac{4}{2} = 2$

A line through the origin with gradient $m$ has equation $y = mx$.

**WATCH OUT**

A gradient can be positive or negative.

+ve up    −ve down

# Gradients

## Worked example

1. Work out the gradient of this line.

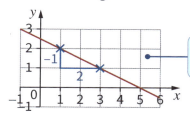

Find two points where you can read the exact coordinates.

$$\text{gradient} = \frac{\text{change in } y}{\text{change in } x} = \frac{-1}{2} = -0.5$$

# Equation of a straight line

The **equation of a straight line** is a rule that is true for all points on the straight line. For example, $y = 2x + 5$.

Equations such as $y = x$ and $y = 2x + 5$ are sloping lines.

The general equation of a straight line is $y = mx + c$

- gradient
- intercept

A **table of values** is a way of writing down the values that satisfy a function or equation. For example:

for $y = 2x$:

| $x$ | 0 | 1 | 2 | 3 |
|---|---|---|---|---|
| $y$ | 0 | 2 | 4 | 6 |

for $y = 2x + 5$:

| $x$ | 0 | 1 | 2 | 3 |
|---|---|---|---|---|
| $y$ | 5 | 7 | 9 | 11 |

## Worked example

1. Complete the table of values for $y = 2x - 3$

| $x$ | −2 | −1 | 0 | 1 |
|---|---|---|---|---|
| $y$ | | −5 | | |

$y = 2 \times (-2) - 3 = -7$
$y = 2 \times 0 - 3 = -3$
$y = 2 \times 1 - 3 = -1$

Substitute the values of $x$ into the equation to find $y$

| $x$ | −2 | −1 | 0 | 1 |
|---|---|---|---|---|
| $y$ | −7 | −5 | −3 | −1 |

2. Draw the graph of $y = 2x - 3$

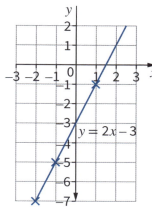

Plot each of the points from the table $(-2, -7), (-1, -5),$ $(0, -3), (1, -1)$

Use a ruler to draw a straight line through all the points.

**LINK**

To remind yourself about rearranging equations, look back at Chapter 5.

**EXAM TIP**

When drawing graphs:
- use a ruler
- check the scales on both axes
- the line must go all the way to the edge of the coordinate grid.

6 Knowledge  43

# ⚙ Knowledge

## 6 Linear graphs

### Equation of a straight line

**Worked example**

**1.** The equation of a straight line is $2y + 6 = 4x$

Work out the gradient and the $y$-intercept of the line.

$2y + 6 - 6 = 4x - 6$ —— Rearrange the equation so it's in the form $y = mx + c$

$2y = 4x - 6$

$\dfrac{2y}{2} = \dfrac{4x - 6}{2}$

$y = 2x - 3$

The gradient ($m$) is 2 and the $y$-intercept ($c$) is $-3$ —— Compare to $y = mx + c$

**2.** Work out the equation of the line.

gradient $(m) = \dfrac{\text{change in } y}{\text{change in } x} = \dfrac{3}{1} = 3$

$y$-intercept $(c) = 4$

$y = mx + c$

$y = 3x + 4$

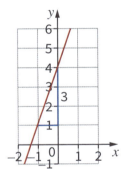

**3.** A line passes through the coordinates (1, 2) and (4, −7).

Work out the equation of the line.

gradient $(m) = \dfrac{\text{change in } y}{\text{change in } x} = \dfrac{-9}{3} = -3$

$y = mx + c \Rightarrow y = -3x + c$

$y = -3x + c$

$2 = -3 \times 1 + c$ —— To find $c$, substitute the coordinate pair (1, 2) into the equation (you could also have used the point (4, −7)).

$2 = 3 + c$

$c = 5$

The equation is $y = -3x + 5$

### Parallel and perpendicular lines

Two lines are **parallel** when they are always the same distance from each other.

A pair of parallel lines will have the same gradient.

$y = 3x + 1$

$y = 3x - 2$

These lines are parallel. They both have a gradient of 3

If two lines are **perpendicular**, the gradient of one line is the negative reciprocal of the other line.

gradient $= m$

gradient $= -\dfrac{1}{m}$

**Key terms** Make sure you can write a definition for these key terms

axes    coordinate    coordinate grid
equation of a straight line    gradient    graph    intercept
linear    origin    parallel    perpendicular    table of terms

## Parallel and perpendicular lines

**Worked example**

A line, $L$, has equation $2y - x + 2 = 0$

Work out the equation of the line that is perpendicular to $L$ and passes through the point $(5, -4)$

$2y - x + 2 = 0$

$2y = x - 2$ — Rearrange into the form $y = mx + c$

$y = \frac{1}{2}x - 1$

gradient of $L = \frac{1}{2}$ — Identify gradient of $L$

negative reciprocal of $\frac{1}{2} = -2$ — Find gradient of the line perpendicular to $L$.

$y = -2x + c$ — Write the equation of the line.

$-4 = -2 \times 5 + c$

$-4 = -10 + c$ — Find the value of $c$ for the perpendicular line by substituting the given coordinates into the equation.

$c = 6$

The equation is $y = -2x + 6$

## Finding approximate values from a linear graph

**Worked example**

Use the graph to find:

**(a)** the value of $x$ when $y = 2$ — Draw a line to the graph at $y = 2$ and read off the $x$-value.

$x = 2.5$

**(b)** the value of $y$ when $x = 2$ — Draw a line to the graph at $x = 1$ and read off the $y$-value.

$y = -1$

## Midpoints of lines

The midpoint of a line segment joining $(x_1, y_1)$ and $(x_2, y_2)$ is

$$\frac{(x_1 + x_2)}{2}, \frac{(y_1 + y_2)}{2}$$

**Worked example**

The straight line $L_1$ passes through the points with coordinates $(4, 7)$ and $(6, 3)$.

**(a)** Work out the midpoint of the line $L_1$

$$\text{midpoint} = \left(\frac{4+6}{2}, \frac{7+3}{2}\right)$$ — Substitute the given coordinates into the formula.

$$= (5, 5)$$

**(b)** The straight line $L_2$ passes through the origin and has gradient $\frac{1}{2}$

The lines $L_1$ and $L_2$ intersect at the point $P$.

Work out the coordinates of $P$.

$L_1$: gradient $= \frac{3-7}{6-4} = -2$

$y = -2x + c$

— Write the equation of $L_1$ by finding the gradient and $y$-intercept using information in the question.

$L_1$ passes through $(6, 3)$.

$3 = -2 \times 6 + c$

$c = 15$

$L_1$: $y = -2x + 15$ — Substitute in $y = 3$ and $x = 6$.

$L_2$ has equation $y = \frac{1}{2}x$

($y$-intercept is 0 as it passes through the origin) — Write the equation of $L_2$

$y = -2x + 15$ ①

$y = \frac{1}{2}x$ ②

Eliminate $y$: $-2x + 15 = \frac{1}{2}x$

$-4x + 30 = x$

$30 = 5x$

$x = 6$

Sub $x = 6$ into ②: $y = \frac{1}{2} \times 6$

$= 3$

— To find where the lines intersect, write them as simultaneous equations and solve.

So the point of intersection is $(6, 3)$

# Retrieval

## 6 Linear graphs

Learn the answers to the questions below, then cover the answers column with a piece of paper and write as many as you can. Check and repeat.

| | Questions | Answers |
|---|---|---|
| 1 | In which directions do the $x$-axis and $y$-axis go? | The $x$-axis is horizontal and the $y$-axis is vertical. |
| 2 | What is a linear graph? | A linear graph is a straight line which can be horizontal, vertical or sloping. |
| 3 | What would the line $x = 4$ look like? | A vertical line passing through $x = 4$. |
| 4 | What would $y = -3$ look like? | A horizontal line passing through $y = -3$. |
| 5 | What is the general equation of a straight line? | $y = mx + c$ |
| 6 | What is a first step to draw a linear graph when you know the equation? | Complete a table of values. |
| 7 | What does the gradient of the line measure? | How steep it is. |
| 8 | If a gradient is negative, what do you know about the line? | It goes downhill from left to right. |
| 9 | How do you calculate the gradient of a line? | $\dfrac{\text{change in } y}{\text{change in } x}$ |
| 10 | What does $m$ stand for in $y = mx + c$? | Gradient. |
| 11 | What does $c$ stand for in $y = mx + c$? | $y$-intercept |
| 12 | What can you say about the gradients of parallel lines? | They are the same. |
| 13 | How would you find the gradient of a line in the following form: $ax + by = c$ (where $a, b, c$ are integers) | Rearrange into the form $y = mx + c$. |

*Put paper here*

### Previous questions

Now go back and use these questions to check your knowledge of previous topics.

| | Questions | Answers |
|---|---|---|
| 1 | What does commutative mean? | A calculation can be done in any order. |
| 2 | When solving equations using the balance method, what do we do first if the unknown is on both sides? | Get the unknown on one side. |
| 3 | What is the opposite of factorising? | Expanding. |

*Put paper here*

## Exam-style questions

**6.1** The straight line $L_1$ passes through the points $(-1, -3)$ and $(4, 2)$

The straight line $L_2$ passes through $(-2, -7)$ and has gradient 4.

The lines $L_1$ and $L_2$ intersect at the point $T$.

Work out the coordinates of $T$. **[4 marks]**

**6.2** On the grid below, draw the graph of $x + y = 5$

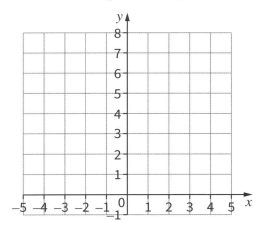

**[3 marks]**

**6.3** On the grid below, draw the graph of $y = 3 - 2x$

**[3 marks]**

# Exam-style questions

 **6.4** Work out the gradient and $y$-intercept of each of these lines.

    **(a)**   $y = 5x + 1$                                    **[2 marks]**

    **(b)**   $y = 3 - 2x$                                **[2 marks]**

    **(c)**   $2y = x + 6$                               **[2 marks]**

    **(d)**   $y - x = 10$                             **[2 marks]**

> **EXAM TIP** 🎯
>
> Rearrange the equations to make $y$ the subject.

 **6.5** Work out the equation of the line parallel to $y = 4x - 8$ that passes through the point $(0, 5)$.             **[1 mark]**

 **6.6** Work out the midpoint of the line segment that joins points $(2, 6)$ and $(4, 7)$.            **[2 marks]**

 **6.7** Work out the equation of the line that passes through the two points.

    **(a)**   $(0, -1)$ and $(2, 3)$                      **[3 marks]**

    **(b)**   $(-3, 5)$ and $(1, 1)$                      **[3 marks]**

 **6.8** Match the equations of a line on the left with the equations of a perpendicular line on the right.

| | |
|---|---|
| $y = \dfrac{1}{2}x + 2$ | $2y = 3x - 4$ |
| $2x = 6 - 3y$ | $y = x - 2$ |
| $x + 2y - 1 = 0$ | $y = -2x + 1$ |
| $y + x = \dfrac{1}{2}$ | $y - 2x = 0$ |

                                                          **[3 marks]**

 **6.9** The straight line $L$ has equation $3x - 6y + 1 = 0$.

    Point $P$ has coordinates $(4, -2)$.

    Work out the equation of the straight line that is perpendicular to $L$ and passes through $P$.             **[4 marks]**

**6.10** Point $A$ has coordinates $(2, 1)$

Point $B$ has coordinates $(p, q)$

A line perpendicular to $AB$ has equation $2y + x = -1$

Work out an expression for $q$ in terms of $p$. **[5 marks]**

 **6.11** $ABC$ is a right-angled triangle. $L$ and $M$ are straight lines.

$L$ has equation $x = 3$.

$CO = \frac{1}{2} OB$

$CB = \frac{4}{3} BA$

Find an equation for $M$.

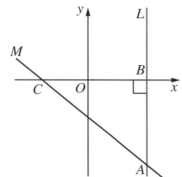

**[5 marks]**

**Questions referring to previous content**

 **6.12** Solve $3 + \dfrac{2(x + 1)}{5} - x = 7 - (2 + x)$ **[3 marks]**

 **6.13** Find the value of at $\dfrac{a - \sqrt{bc}}{c^2 - b}$ when $a = 10$, $b = -18$, $c = -2$ **[4 marks]**

## 7 Real-life graphs

## Finding approximate values from a linear graph

> **Worked example**
>
> 1. Find the value of $x$ when $y = 2$.
>
>    $x = 2.5$
>
> 2. Find the value of $y$ when $x = 1$.
>
>    $y = -1$

> Draw a line to the graph at $y = 2$ and read off the $x$ value.

> Draw a line to the graph at $x = 1$ and read off the $y$ value.

## Real-life graphs

**Real-life graphs** are used to represent real-life situations and data. They can be straight lines or curves.

> **Worked example**
>
> The graph shows the cost of Peter's phone contract per month.
>
> 1. Explain why the graph starts at (0, 10).
>
>    For 0 hours, Peter pays £10. This is a fixed cost before he makes any calls.
>
> 2. One month, Peter makes 40 hours of calls. Work out how much his bill is for that month.
>
>    > Look to see where the line is when the $x$-axis is at 40 hours.
>
>    £30
>
> 3. Work out how much a call costs per hour.
>
>    > You need to find the **gradient**.
>
>    £5 ÷ 10 hours = £0.50 per hour of calls.

> **LINK**
>
> To remind yourself about gradients, look back at Chapter 6.

> **Formula**
>
> $$\text{gradient} = \frac{\text{change in } y}{\text{change in } x}$$

> **WATCH OUT** !
>
> Always check what each unit on each scale represents.

# Distance-time graphs

A distance-time graph shows how the distance changes as time passes.
On a distance-time graph, gradient = **speed**.

$$gradient = \frac{distance}{time}$$
$$= \textbf{speed}$$

**A.** Straight sloped line means the speed is constant.

**B.** Horizontal section means time is passing but the distance isn't changing, so the speed is 0.

**C.** Steepest section, so the speed was the quickest here.

---

**Worked example**

The graph shows the height of a chilli plant at the end of each week.

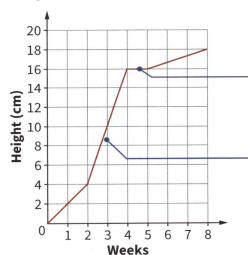

The line is flat between 4 and 5 weeks, which means the plant didn't increase in height in this time.

The line is steepest between 2 and 4 weeks, this means the height was increasing faster than at any other time.

1. How tall was the plant after three weeks?

   It was 10 cm tall.

2. In which week did the plant not grow?

   It did not grow between weeks 4 and 5.

3. When did the plant grow the fastest and what was its speed of growth during this time?

   The fastest increase was between 2 and 4 weeks. The speed of growth was $\frac{12}{2} = 6$ cm/week.

### Distance-time graphs

---

**Worked example**

Min-Su drives a total of 14 km to work.

- She travels the first 6 km at a constant speed of 36 km/h.
- She then stops for 5 minutes at roadworks.
- It takes her 15 minutes to complete the rest of the journey at a constant speed.

Draw a distance-time graph of the journey.

Work out the time taken to travel first 6 km:

| distance | time |
|----------|------|
| 36 km    | 1hr  |
| 6 km     | $\frac{1}{6}$ hr |

÷6 ... ÷6

$\frac{1}{6}$ hr = $\frac{60}{6}$

= 10 min

5 min stop.

Total journey = 14 km.

15 mins constant speed.

**WATCH OUT** ❗

Be careful of the units – check if the time is given in minutes or hours.

---

**Worked example**

The distance-time graph shows how a snail moved one morning.

What was its speed in m/h between 09:00 and 09:30?

The snail moved 4 m in 0.5 hour.

| distance | time   |
|----------|--------|
| 4 m      | 0.5 hr |
| 8 m      | 1 hr   |

×2 ... ×2

speed = 8 m/hr ●——— Use hours, not minutes here.

**Key terms**

Make sure you can write a definition for these key terms.

*acceleration*
*real-life graph*
*speed*

# Speed–time graphs

A speed-time graph shows how the speed changes as time passes.

The gradient of a speed-time graph is the equal to the **acceleration**.

$$\text{gradient} = \frac{\text{speed}}{\text{time}}$$

$$= \textbf{acceleration}$$

**A.** Straight sloped lines mean that acceleration is constant.

**B.** Horizontal section means time is passing but the speed isn't changing, so the speed is constant.

**C.** Negative gradient, which means deceleration or slowing down.

**D.** Speed = 0; they have stopped.
The area under the graph represents the distance travelled.

# Rates of change

## Worked example

These vases are filled with water at a constant rate. Sketch a graph for each to show how the depth of water varies over time.

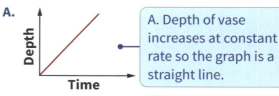

A. Depth of vase increases at constant rate so the graph is a straight line.

B. Rate of change of vase depth decreases as it gets wider, so the graph is a curve.

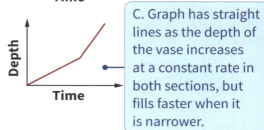

C. Graph has straight lines as the depth of the vase increases at a constant rate in both sections, but fills faster when it is narrower.

**LINK**
You will revise compound measures in Chapter 14.

## Worked example

The graph shows the speed of a runner at the start of a 100 m race.

**(a)** Find their initial acceleration in m/s.
2.5 m/s

**(b)** Find the distance covered by the runner in the first 4 seconds.

$$\text{area} = \frac{1}{2} \times \text{base} \times \text{height}$$

$$= \frac{1}{2} \times 4 \times 10 = 20 \text{ m}.$$

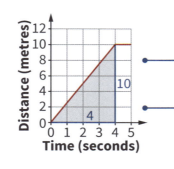

$$\text{gradient} = \frac{\text{change in } y}{\text{change in } x}$$

$$= \frac{10}{4}$$

$$= 2.5 \text{ m/s}$$

The distance travelled is the area under the graph between 0 and 4 seconds.

This is a triangle with a base of length 4 and a height of 10.

# ⮂ Retrieval

## 7 Real-life graphs

Learn the answers to the questions below, then cover the answers column with a piece of paper and write as many as you can. Check and repeat.

| | Questions | | Answers |
|---|---|---|---|
| 1 | On a speed–time graph, what does the gradient represent? | | Acceleration. |
| 2 | What does a straight line represent on a distance–time graph? | | Constant speed. |
| 3 | On a distance–time graph, what does a horizontal section represent? | | No movement; the speed is 0. |
| 4 | On a speed–time graph, what does a horizontal section represent? | | Constant speed; the acceleration is 0. |
| 5 | How do you calculate speed from distance and time? | | $speed = \dfrac{distance}{time}$ |
| 6 | Why can't distance–time graphs have negative values? | | You can't go back in time or have a negative distance. |
| 7 | How do you calculate distance from speed and time? | | $distance = speed \times time$ |
| 8 | What does a negative gradient mean on a speed–time graph? | | The object is slowing down (decelerating). |

*Put paper here*

## Previous questions

Now go back and use these questions to check your knowledge of previous topics.

| | Questions | | Answers |
|---|---|---|---|
| 1 | In a sale, a hat is reduced from \$34 to \$28. Find the percentage decrease. | | $\% \text{ decrease} = \dfrac{\text{actual decrease}}{\text{original amount}} \times 100\%$ $= \dfrac{34 - 28}{34} \times 100\%$ $= 17.6\%$ |
| 2 | What is standard form used for? | | To make it easier to understand very large or very small numbers. |
| 3 | Write $a^3 \div a^{-5}$ as a single power of $a$. | | $a^3 \div a^{-5} = a^{3-(-5)}$ $= a^{3+5}$ $= a^8$ |

*Put paper here*

## Exam-style questions

 **7.1** Krystyna runs a race.

Here is her velocity-time graph.

**(a)** Work out Krystyna's acceleration during the first 15 seconds.

State the units of your answer. **[2 marks]**

**(b)** Work out the distance travelled by Krystina during the race. **[3 marks]**

**7.2** Kai cycles from his home.

Here is part of a distance–time graph showing part of Kai's journey.

**(a)** Work out Kai's average speed in the first hour and a half of his ride.

Give your answer in km/h. **[3 marks]**

**(b)** At 12:30, Kai stops for lunch for 45 minutes before cycling back home at 20 km/h.

Complete the distance–time graph. **[2 marks]**

**7.3** The graph shows the costs of taxi journey for different distances.

**Distance (miles)**

(a) Interpret the meaning of the y-intercept. [1 mark]

(b) Interpret the meaning of the gradient. [1 mark]

(c) How much would a 12-mile journey cost? [2 marks]

(d) State any assumptions that you have made in part your answer to part (c). [1 mark]

 **7.4** The graph shows the height of garden plant A over a period of 6 weeks.

**Number of weeks**

(a) In which week was there no growth? [1 mark]

(b) When was the growth fastest? [1 mark]

(c) Garden plant B grows at an average rate of 8 cm per week. Which garden plant grows slowest? Show your working. [2 marks]

**Questions referring to previous content**

 **7.5** The point with coordinates $(-2, -1)$ lies on line $L_1$ with equation $y - 5 - 3x = 0$

The line $L_2$ also passes through $(-2, -1)$ and is perpendicular to L1.

Find the $x$-coordinate of the point on with $y$-coordinate $-3$. [4 marks]

 **7.6** Which two lines are perpendicular to each other? [2 marks]

$-2x + y = -2$    $5y + 2x = 10$    $\frac{x}{3} + y = -2$    $-2x + \frac{4}{5}y = -10$

## 8 Solving inequalities in 1 or 2 variables

### Using inequalities

When two expressions are not equal, you can use an **inequality** to show which is bigger.

For example: $3x + 15 \geq 25$

**REVISION TIP**

Remember, the small end of the inequality sign points to the smaller value. For example, $5 < 9$ and $15 > 6$

When an inequality contains an **unknown** $x$, you can solve it to find the values of $x$ that **satisfy** the inequality.

**Worked example**

Solve $2x + 13 \geq 25$

$$2x + 13 \geq 25$$
$$2x + 13 - 13 \geq 25 - 13$$
$$2x \geq 12$$ — Rearrange the terms so you have numbers on one side and $x$ on the other.

$$\frac{2x}{2} \geq \frac{12}{2}$$ — Divide both sides by the multiple of $x$.

$$x \geq 6$$ — Write the solution as an inequality.

**WATCH OUT**

As with equations, make sure you do the same operation to both sides of the inequality.

If you multiply or divide both sides of an inequality by a negative number, the direction of the inequality changes.

**Worked example**

Solve $-2x + 13 \geq 25$

**Method 1:** Dividing by a negative and reversing the sign of the inequality.

$$-2x + 13 \geq 25$$ — Rearrange the terms so you have numbers on one side and $x$ on the other.
$$-2x \geq 25 - 13$$
$$-2x \geq 12$$

$$\frac{-2x}{-2} \geq \frac{12}{-2}$$ — Divide both sides by the multiple of $x$.
$$x \leq -6$$

Write the solution as an inequality.

**WATCH OUT** ❗

Remember to change the direction of the inequality because it was divided by a negative number.

**Method 2:** Rearranging the equation so you only have to divide by a positive.

$$-2x + 13 \geq 25$$ — Take the $x$ term to the other side of the equation to make it positive.
$$13 \geq 25 - 2x$$
$$13 - 25 \geq 2x$$
$$-12 \geq 2x$$ — Now take the number to the other side and simplify.

$$-\frac{12}{2} \geq \frac{2x}{2}$$ — Divide both sides by the multiple of $x$.
$$-6 \geq x$$

Write the solution as an inequality.

**WATCH OUT** ❗

Inequality signs don't change when terms are divided by a positive.

You could be given two inequalities and asked to find values of $x$ which satisfy both inequalities.

A **number line** can be used to represent the values of $x$ that are included in the solution to the inequality.

A hollow circle shows that the number is not included (less/greater than).

A filled-in circle shows that the number is included (less/greater than or equal to).

# ⚙ Knowledge

## 8 Solving inequalities in 1 or 2 variables

## Using inequalities

**Worked example**

$2x + 1 > 7$ and $x - 4 \leq 5$

What values of $x$ satisfy both inequalities? Show your answer on a number line.

| $2x + 1 > 7$ | $x - 4 \leq 5$ | Solve each inequality. |
|---|---|---|
| $2x > 6$ | $x \leq 9$ | |
| $x > 3$ | $3 < x \leq 9$ | Combine the inequalities. Take care to write the inequality sign the right way. |

Draw on a number line.

## Representing inequalities on graphs

Solutions to a set of inequalities can be represented on a graph.

- **strict inequalities** ($<$ and $>$) are represented with a dashed line.
- **non-strict inequalities** ($\leq$ and $\geq$) are represented with a solid line.

**Worked example**

Write an inequality to satisfy the shaded region.

Identify the gradient and $y$-intercept to work out the equation of the line.

$m = -1$ and $c = 2$

so $y = -1x + 2$
$\quad = 2 - x$

$y < 2 - x$

$0 < 2 - 0$ ✓

Substitute $(0,0)$ into the inequality to check it.

Identify the inequality:

- A dashed line means it is not included in the inequality.
- The shaded area is below the line, so the inequality is 'less than'.

**Worked example**

On the grid, identify the region rerpresented by $y > 4 - 2x$, $x \leq 7$, and $y \leq 3$

Label the region $R$.

Draw the lines $y = 4 - 2x$, $x = 7$ and $y = 3$.

Draw a dashed line for $y = 4 - 2x$ because it is a strict inequality.

Consider each inequality in turn:

- $x \leq 7$, so you need $x$-values smaller than or equal to 7.
- $y \leq 3$, so you need $y$-values smaller than or equal to 3.
- For $y > 4 - 2x$, the inequality is 'greater than' so the shaded area is above the line.

Shade the area that satisfies all the inequalities.

 **Key terms** — **Make sure you can write a definition for these key terms**

expression   inequality
non-strict inequality
strict inequality   number line
satisfy   solve   unknown

## 8 Solving inequalities in 1 or 2 variables

Learn the answers to the questions below, then cover the answers column with a piece of paper and write as many as you can. Check and repeat.

| | Questions | Answers |
|---|---|---|
| 1 | What is an inequality? | A relationship between two expressions or values that are not equal to each other. |
| 2 | How do you represent an inequality on a number line where the value is **not** included? | Draw an open circle. |
| 3 | What should you do when you multiply or divide an inequality by a negative number? | Change the direction of the inequality symbol. |
| 4 | What does a solid circle mean when an inequality is represented on a number line? | The number below the circle is included in the inequality. |
| 5 | What does the $a > b$ mean? | $a$ is greater then $b$. |
| 6 | What does the $a \leq b$ mean? | $a$ is less than or equal to $b$. |
| 7 | How do you solve two inequalities that require you to show the solutions on a number line? | Solve each inequality separately then combine the solutions. |
| 8 | On a graph, would you use a dotted or solid line to represent the inequality $2x > 5$? | Dotted, as the points on the line do not satisfy the inequality. |
| 9 | On a graph, would you use a dotted or solid line to represent the inequality $6 \geq 5x$? | Solid, as the points on the line satisfy the inequality. |

*Put paper here*

## Previous questions

Now go back and use these questions to check your knowledge of previous topics.

| | Questions | Answers |
|---|---|---|
| 1 | How do you convert a decimal into a percentage? | Multiply by 100. |
| 2 | What does truncate mean? | Remove digits without rounding. |
| 3 | What does it mean to expand a single bracket? | To multiply each term inside the bracket by the term in front of the bracket. |
| 4 | When a number is written in standard form, what form does it take? | $a \times 10^n$, where $1 <= a < 10$ and $n$ is an integer. |
| 5 | How do you multiply fractions? | Multiply the numerators and multiply the denominators. |

*Put paper here*

# Practice

## Exam-style questions

**8.1** Solve these inequalities and show the solutions on the number lines.

**EXAM TIP**

Remember to include the inequality symbol in your final answers.

**(a)** $3x + 5 > 2$

[3 marks]

**(b)** $20 - 5x \geq 0$

[3 marks]

**8.2** **(a)** Solve $-6 < x - 4 \leq -1$ [2 marks]

**(b)** Show the solution on the number line.

[1 mark]

**8.3** Work out the set of values of $x$ that satisfy both inequalities.

$4x + 1 \geq 3$

$5 - x > 3$ [3 marks]

**8.4** Write the inequality that describes each of these shaded regions.

**(a)**

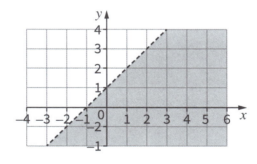

[1 mark]

**(b)**

[1 mark]

**8.5** On the grid below, identify the region that represents.

$y < 2x \quad y \geq -2 \quad x + y \geq 1$

Label the region $R$.

**[3 marks]**

 **8.6** **(a)** Write the three inequalities that define the shaded region. **[3 marks]**

**(b)** Write the points in the region with integer coordinates. **[1 mark]**

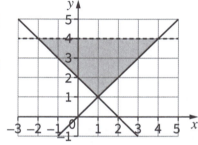

**8.7** I think of a number.

If I halve my number and subtract 4, I get more than if I double my number and add 3.

Work out the possible values for my number. **[4 marks]**

**8.8** The width of a rectangle is 5 cm. The length is $x$ cm longer than the width.

The perimeter of the rectangle is less than 20 cm.

Form and inequality and solve it to find to find the range of possible values for $x$. **[4 marks]**

**Questions referring to previous content**

 **8.9** Draw a distance-time graph to represent Alex's journey to school. **[5 marks]**

She travels for 25 minutes at 24 km/h.

She then gets stuck in traffic for 9 minutes

After the traffic clears, she drives the remaining 15 km in 11 minutes.

Use your graph to find Alex's average speed over the whole journey to school.

 **8.10** Which one of the following is the only true statement about speed-time graphs? **[1 mark]**

A line segment with zero gradient means the object is at rest.

B A line segment with negative gradient means the object is moving backwards.

C An upwards-sloping line segment represents constant acceleration.

D The maximum point on the graph represents maximum acceleration.

# Knowledge

## 9 Factorising quadratic expressions and solving quadratic equations

## Quadratic expressions

A **quadratic expression** is an expression where the highest power of the variable is 2.

They are of the form $ax^2 + bx + c$, where $a$, $b$ and $c$ are the **coefficients**.

For example, $x^2 + 5x - 7$ is a quadratic expression, where $a = 1$, $b = 5$, and $c = -7$.

If the coefficient of $b$ is 0, the expression is: $x^2 - c$

$c$ will be a square number.

If the coefficient of $c$ is 0, the expression is: $x^2 + bx$

## Difference of two squares

The **difference of two squares** is an expression, such as $a^2 - b^2$, which contains one square number minus another square number. For example, $x^2 - 9$ is the difference of two squares.

### Formula

The difference of two squares can be factorised:

$a^2 - b^2 = (a + b)(a - b)$

When you multiply out the brackets, the middle terms $-ab$ and $+ab$ cancel out and leave just $a^2 - b^2$

| × | $a$ | $+b$ |
|---|-----|------|
| $a$ | $a^2$ | $+ab$ |
| $-b$ | $-ab$ | $-b^2$ |

$a^2 + ab - ab - b^2$
$= a^2 - b^2$

### Worked example

Factorise $4x^2 - 9$

$4x^2 = (2x)^2$ and $9 = 3^2$,

so, $4x^2 - 9 = (2x)^2 - 3^2$

$\qquad = (2x - 3)(2x + 3)$

$\qquad = 4x^2 - 6x + 6x - 9$ •——

$\qquad = 4x^2 - 9$

Check by multiplying out.

## Expanding double brackets

To **expand** (multiply out) a pair of brackets, multiply each term in the first set of brackets by each term in the second pair of brackets.

$(x + 5)(2x - 7) = 2x^2 - 7x + 10x - 35$

$\qquad\qquad = 2x^2 + 3x - 35$ •— Collect like terms.

If you prefer, you can use a multiplication grid.

| × | $x$ | 5 |
|---|-----|-----|
| $2x$ | $2x^2$ | $10x$ |
| $-7$ | $-7x$ | $-35$ |

## Solving quadratic equations by factorising

You can solve some quadratic equations by **factorising**.

### Worked example

**Solve** $x^2 + 6x = 7$

$x^2 + 6x - 7 = 0$ •— Rearrange into the form $ax^2 + bx + c = 0$

$x^2 + 6x - 7x - 7 = 0$

$x(x + 6) - 7(x + 6) = 0$ •—

$(x - 7)(x + 6) = 0$

$x + 7 = 0$ •—

$x - 1 = 0$

$x = -7$

$x = 1$

Factorise into double brackets by finding two numbers that add to give 6 and multiply to give $1 \times -7 = -7$

Numbers are 7 and −1:

$7 + (-1) = 6$

$7 \times (-1) = -7$

Solve each new equation and write the values of $x$.

Set each factor equal to zero.

# Factorising quadratic equations

**Factorising** is the opposite of expanding.

Some expressions in the form $ax^2 + bx + c$ can be factorised into a double set of brackets.
For example: $x^2 + 6x + 5 = (x + 5)(x + 1)$

**Worked example**

Factorise $6x^2 + 7x - 3$

$a = 6, b = 7, c = -3$

$6 \times -3 = -18$ — Multiply coefficients $a$ and $c$.

$9 + -2 = 7$
$9 \times -2 = -18$ — Find two numbers whose product is $ac$ (= −18) and whose sum is $b$ (= 7). The numbers are 9 and −2.

$6x^2 + 9x - 2x - 3$ — Rewrite the quadratic expression using 9 and −2.

$= (6x^2 + 9x) + (-2x - 3)$ — Group the terms.

$= 3x(2x + 3) - 1(2x + 3)$ — Factor out the HCF from each group.

$= (3x - 1)(2x + 3)$ — Factor out the terms within the parentheses that are the same, $(2x + 3)$.

**REVISION TIP**

Expand the brackets to check that you factorised correctly.

Quadratic expressions that have two terms, and a common factor, e.g. $ax^2 + bx$ can be factorised into single brackets. For example: $x^2 + 3x = x(x + 3)$

**Worked example**

Factorise $5x^2 - 15x$

$HCF = 5x$ — Find the HCF of $5x^2$ and $15x$.

$\dfrac{5x^2}{5x} - \dfrac{15x}{5x}$ — Divide each term in the expression by the HCF.

$= x - 3$

$= 5x(x - 3)$ — Write the HCF outside the brackets, and write the divided terms inside the bracket.

**WATCH OUT**

Make sure you use the same sign inside the bracket as is in the original expression.

# Knowledge

## 9 Factorising quadratic expressions and solving quadratic equations

### Solving quadratic equations by completing the square

Quadratic equations that can't be solved by factorising can be solved by 'completing the square'. The factorised equation takes the form $(x + p)^2 + q$, where $p$ and $q$ are integers.

**Worked example**

By completing the square, write $x^2 + 6x + 17$ in the form $(x + p)^2 + q$, where $p$ and $q$ are integers.

17 is a prime number, so it isn't possible to factorise.

Write the expression without the constant term.

Halve the coefficient of $x$ to give you the constant term in the squared bracket. Replace the numerical term with $-c$ to retain the balance of the original equation.

Since we have a term that is being squared, we can picture this as a square that has sides of length $x + 3$.

**REVISION TIP**

We can also find $c$ by expanding the brackets.

$(x + 3)^2$

$= x^2 + 3x + 3x + 9$

$= x^2 + 6x + 9$

Remember to use the value of $-c$ to complete the square.

$= (x + 3)^2 - 9$

$x^2 + 6x$
$= (x + 3)^2 - c$

$x + 3$

$x$      $3$

This square has an area of $x^2$.

Each side of the square has length $x + 3$

$x$ | $x^2$ | $3x$

These two rectangles have a combined area of $6x$.

$x + 3$

$3$ | $3x$ | $3^2 = 9$ | $3$

The large square has area $(x + 3)^2$

The missing area is $3^2 = 9$

$3$

The square of area 9 is missing from the expression, so to complete the square, subtract this from the factorized expression.

Add the constant term from the original expression into the factorized expression.

Simplify.

$= (x + 3)^2 - 9$
$= (x + 3)^2 - 9 + 17$
$= (x + 3)^2 + 8$

## Solving quadratic equations by completing the square

Quadratic equations may need to be rearranged before you can solve them.

**Worked example**

Solve $2x^2 = 8x + 14$. Give your answer in the form $a \pm b\sqrt{c}$, where $a$, $b$ and $c$ are integers.

$$2x^2 = 8x + 14$$ — Divide through by 2 so the coefficient of $x^2$ is 1

$$x^2 = 4x + 7$$

$$x^2 - 4x - 7 = 0$$ — Write as a quadratic expression equal to 0

$$x^2 - 4x = (x-2)^2 - k \text{ for some value of } k$$ — Complete the square.

$$x^2 - 4x = x^2 - 4x + 4 - k$$

$$\text{so } k = 4$$

$$x^2 - 4x = (x-2)^2 - 4$$

So $x^2 - 4x - 7 = 0$ becomes:

$$(x-2)^2 - 4 - 7 = 0$$

$$(x-2)^2 - 11 = 0$$

$$(x-2)^2 = 11$$

$$x - 2 = \pm\sqrt{11}$$ — Take the square root of both sides and rearrange to make $x$ the subject.

$$x = 2 \pm \sqrt{11}$$

## Solving quadratic equations using the quadratic formula

The method for completing the square for the general quadratic equation $ax^2 + bx + c = 0$, gives the quadratic formula:

$$x = \frac{-b \pm \sqrt{b^2 - 4ac}}{2a}$$

**Worked example**

Solve $5x^2 + 7x - 3 = 0$

$a = 5$, $b = 7$, $c = -3$ — Identify the values of $a$, $b$ and $c$.

$$x = \frac{-7 \pm \sqrt{7^2 - 4 \times 5 \times (-3)}}{2 \times 5}$$ — Substitute these values into the quadratic formula.

$$x = \frac{-7 \pm \sqrt{109}}{10}$$

$$x = \frac{-7 + \sqrt{109}}{10} \quad x = \frac{-7 - \sqrt{109}}{10}$$ — Solve to find the two possible values for $x$.

$x = 0.344$ $\quad x = -1.74$ (to 3 s.f.)

**Key terms** Make sure you can write a definition for these key terms.

coefficient   difference of two squares   expand
factor   factorise   quadratic expression   solve

**9 Knowledge** 65

## 9 Factorising quadratic expressions and solving quadratic equations

Learn the answers to the questions below, then cover the answers column with a piece of paper and write as many as you can. Check and repeat.

| | Questions | Answers |
|---|---|---|
| 1 | What is a quadratic equation? | An equation where the highest power of the variable is 2. |
| 2 | What should be your first step if you want to solve $x^2 + x = 12$? | Subtract 12 from both sides so the equation is equal to 0. |
| 3 | If you factorise $x^2 + bx + c$ you need to find two numbers which add to give which letter? | $b$ |
| 4 | If you factorise $x^2 + bx + c$ you need to find two numbers which multiply to give which letter? | $c$ |
| 5 | What is an expression in the form $a^2 - b^2$ called? | The difference of two squares. |
| 6 | What does $a^2 - b^2$ factorise to? | $(a - b)(a + b)$ |
| 7 | What is the general form of a quadratic expression? | $ax^2 + bx + c$ |
| 8 | How do you solve a quadratic expression that can be factorised? | Rearrange the equation such that $ax^2 + bx + c = 0$ then factorise. |
| 9 | When might you use the quadratic equation? | To solve a quadratic equation that cannot be factorised. |
| 10 | What form do you write a quadratic expression in when you complete the square? | $a(x + b)^2 + c$ |

*Put paper here*

## Previous questions

Now go back and use these questions to check your knowledge of previous topics.

| | Questions | Answers |
|---|---|---|
| 1 | What does the letter I in BIDMAS stand for? | Index or indices. |
| 2 | What does $5a$ mean? | 5 multiplied by $a$. |
| 3 | What is the opposite of expanding? | Positive. |
| 4 | What does the gradient of a line tell you about the line? | How steep the line is. |
| 5 | What is rationalising the denominator? | Rewriting a fraction to remove any surds from the denominator. |

*Put paper here*

## Exam-style questions

 **9.1** A cube has side length $(x + 1)$ cm.

Work out an expression for the volume of the cube in the form
$ax^3 + bx^2 + cx + d$.                                            **[3 marks]**

 **9.2** Factorise

**(a)** $p^2 - 9p - 36$                                            **[2 marks]**

**(b)** $x^2 + 9p - 10$                                            **[2 marks]**

 **9.3** A rectangle has area $x^2 + 6x - 27$

Work out expressions for the lengths of the two sides of the rectangle.

$x^2 + 6x - 27$

                                                                  **[2 marks]**

 **9.4** Factorise fully

**(a)** $y^2 - 81$                                                **[1 mark]**

**(b)** $4b^2 - 1$                                                **[1 mark]**

**(c)** $28 - 4y^2$                                               **[1 mark]**

 **9.5** **(a)** Factorise $p^2 - q^2$                      **[1 mark]**

**(b)** Hence simplify fully $(3 - x)^2 - (3x + 3)^2$             **[3 marks]**

 **9.6** Factorise

**(a)** $6x^2 + 5x - 4$                                           **[2 marks]**

**(b)** $12x^2 + 20x + 3$                                         **[2 marks]**

 **9.7** A square has area $9x^2 + 30x + 25$. Work out the perimeter of the
square in terms of $x$.                                           **[2 marks]**

# Exam-style questions

**9.8**  Solve

**(a)**  $x^2 - 7x - 8 = 0$                                              [3 marks]

**(b)**  $x^2 + 5x = 0$                                                   [3 marks]

**(c)**  $x^2 - 1 = 0$                                                     [3 marks]

**9.9**  Solve

**(a)**  $x^2 - 12x = -35$                                            [4 marks]

**(b)**  $x^2 + 3x = 10$                                               [4 marks]

> **EXAM TIP**
>
> Before solving, make sure that one side of the equation is zero.

**9.10**  Here is a rectangle.

The area of the rectangle is $5\,\text{cm}^2$.

**(a)**  Show that $2x^2 - 5x - 3 = 0$                          [3 marks]

**(b)**  Work out the length of the shorter side of the rectangle.                                [3 marks]

$5\,\text{cm}^2$   $(x - 2)\,\text{cm}$

$(2x - 1)\,\text{cm}$

**9.11**  Solve these equations. Give your answers to 3 significant figures where appropriate.

**(a)**  $3x^2 - 7x - 10 = 0$                                       [3 marks]

**(b)**  $5x^2 + 4x - 20 = 0$                                       [3 marks]

**(c)**  $25 + 10x - 2x^2 = 0$                                     [3 marks]

**9.12**  Solve these equations.

Give your answers in the form $a \pm b\sqrt{c}$, where $a$, $b$ and $c$ are integers.

**(a)**  $x^2 - 10x + 7 = 0$                                        [3 marks]

**(b)**  $x^2 + 2x - 11 = 0$                                        [3 marks]

> **EXAM TIP**
>
> You can either use the quadratic formula or completing the square to solve these equations.

**9.13**  Pavel says that $2x^2 + 7x + 6$ factorises to $(2x + 6)(x + 1)$, since the sum of 6 and 1 is 7 and the product of 6 and 1 is 6. Pavel is wrong. Explain why.                                      [2 marks]

**9.14**  Write these expressions in the form $(x + p)^2 + q$

**(a)**  $x^2 + 6x + 5$                                                [2 marks]

**(b)**  $x^2 - 4x + 10$                                              [2 marks]

**(c)**  $x^2 - 10x$                                                    [2 marks]

**(d)**  $x^2 + x + 1$                                                  [2 marks]

**9.15**  **(a)** Complete the square for the expression $x^2 - 2x - 11$ **[2 marks]**

**(b)** Hence, solve the equation $x^2 - 2x - 11 = 0$

Give your answer in the form $a \pm b\sqrt{3}$, where $a$ and $b$ are integers. **[2 marks]**

**9.16**  Write each expression in the form $a(x + p)^2 + q$

**(a)** $2x^2 + 8x + 7$ **[3 marks]**

**(b)** $12 + 10x - x^2$ **[3 marks]**

**9.17**  **(a)** Complete the table of values for $y = x^2 + 5x - 2$

| $x$ | −6 | −5 | −4 | −3 | −2 | −1 | 0 | 1 |
|---|---|---|---|---|---|---|---|---|
| $y$ | | −2 | | | −8 | | | 4 |

**[2 marks]**

**(b)** On the grid, draw the graph of $y = x^2 + 5x - 2$ for values of $x$ from −6 to 1

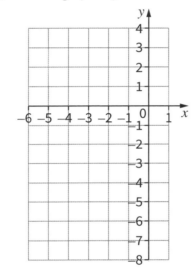

**[2 marks]**

**(c)** Use the graph to estimate a solution to $x^2 + 5x - 2 = 0$ **[1 mark]**

**Questions referring to previous content**

**9.18**  Solve $2(x + 1) - 4 \leq 3\left(x - \frac{4}{3}\right)^3 < 3 + 2x$ **[3 marks]**

**9.19**  On a coordinate grid, sketch the following inequalities and describe the shape that they bound. **[4 marks]**

$$x + y > 4 \qquad 3y - 7x > -28 \qquad 7y - 3x \leq 28$$

# ⚙ Knowledge

## 10 Quadratic graphs, iterations, solving quadratic inequalities

## Quadratic graphs

A **quadratic graph** is one with the general equation:

$$y = ax^2 + bx + c$$

where $a$, $b$, and $c$ are numbers.

**Roots** are the solutions of $ax^2 + bx + c = 0$.
These are the $x$-intercepts of the graph.

The maximum or minimum point of the graph is called the **turning point**.

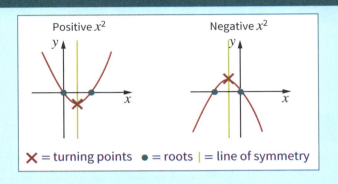

Positive $x^2$      Negative $x^2$

$\times$ = turning points    ● = roots  | = line of symmetry

All quadratic graphs have a **line of symmetry** through the turning point.

Quadratic graphs may have 2, 1 or 0 intercepts.

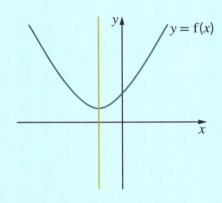

2 $x$-intercepts, so f($x$) has
2 real roots.

1 $x$-intercept, so f($x$) has
1 real root.

0 $x$-intercepts, so f($x$) has
0 real roots.

Roots, intercepts, and turning points can be found using a graph.

---

### Worked example

Here is the graph $y = x^2 + 2x - 1$

1. Write down the coordinates of the turning point of $y = x^2 + 2x - 1$

   $(-1, -2)$

2. Write down estimates for the roots of $x^2 + 2x - 1 = 0$

   Approximately, $x = 2.4$ and $x = 0.4$

The turning point ($\times$) is the minimum point.

The roots (●) are the $x$-intercepts of the graph.

---

# Quadratic graphs

Roots, intercepts, and turning points can also be found algebraically.

**Worked example**

$y = 2x^2 + 11x + 12$

**(a)** Work out the coordinates of the points where the graph intersects the $x$- and $y$-axes.

$y = 2 \times 0^2 + 11 \times 0 = 12,$

$y$-intercept = **(0, 12)**

> The $y$-**intercept** occurs when $x = 0$.
> So substitute 0 for $x$ and solve to find $y$.

$2x^2 + 11x + 12 = 0$

$2x^2 + 3x + 8x + 12 = 0$

$x(2x + 3) + 4(2x + 3) = 0$

$(x + 4)(2x + 3) = 0$

> The $x$-intercepts (roots) occur when $y = 0$.
> So set the equation equal to 0.

> Factorise the equation.

When $x + 4 = 0$, $x = -4$, so **(−4, 0)** is a root.

When $2x + 3 = 0$, $x = -1.5$, so **(−1.5, 0)** is a root.

> Set each factor equal to 0 and solve.

**LINK**

For a reminder of solving quadratic equations, look back at Chapter 9.

**(b)** Find the coordinates of the turning point of the graph.

$y = 2\left[x^2 + \dfrac{11}{2}x + 6\right]$

> Find the lowest possible value for $y$ by completing the square. Take out a factor of 2 (the coefficient of $x^2$).

$= 2\left[\left(x + \dfrac{11}{4}\right)^2 - \left(\dfrac{11^2}{4^2} + 6\right)\right]$

$= 2\left[\left(x + \dfrac{11}{4}\right)^2 - \left(\dfrac{25}{16}\right)\right]$

> Halve the coefficient of $x$ to give the constant term in the squared bracket.

$= 2\left(x + \dfrac{11}{4}\right)^2 - \left(\dfrac{25}{8}\right)$

> Combine the constant terms.

$\dfrac{25}{8} = 2\left(x + \dfrac{11}{4}\right)^2$

> Multiply by 2, then add $\dfrac{25}{8}$.

$\left(x + \dfrac{11}{4}\right)^2 = 0$

$x = -\dfrac{11}{4}$

> Now you have the minimum-possible $y$-value, find the corresponding $x$-value.

Turning point is at $\left(-\dfrac{11}{4}, -\dfrac{25}{8}\right) = (-2.75, -3.125)$

**(c)** Hence, **sketch** the graph of $y = 2x^2 + 11x + 12$

**REVISION TIP**

When you are asked to sketch a graph, you only need to show the shape, the coordinates where the line intercepts the axes, and the turning point.

If you are asked to plot a graph, you need to use a table of values and plot it on graph paper.

## 10 Quadratic graphs, iterations, solving quadratic inequalities

### Quadratic inequalities

Quadratic inequalities can also be solved algebraically.

---

**Worked example**

Solve $2x^2 - 5x \leq 12$, giving your answer in set notation.

$2x^2 - 5x - 12 \leq 0$     → Rearrange to get 0 on its own on one side of the inequality.

$2x^2 - 5x - 12 = 0$

$(2x + 3)(x - 4) = 0$     → Rewrite as an equation and factorise.

$2x + 3 = 0$, so $2x = -3$

  and $x = -\dfrac{3}{2}$

or     → Set each factor equal to 0 and solve for $x$.

$x - 4 = 0$, so $x = 4$

Draw a sketch to help. You are solving $2x^2 - 5x - 12 \leq 0$, so you need the part of the graph where $y \leq 0$ (shown in red).

$-\dfrac{3}{2} \leq x \leq 4$

in set notation: $\left\{ x: -\dfrac{3}{2} \leq x \leq 4 \right\}$     → Give your answer in set notation.

> **REVISION TIP**
>
> When sketching a graph to help solve a quadratic inequality, you only need to label the roots. These tell you the $x$-values for which the graph is positive and negative.

---

**Worked example**

Work out the values of $p$ for which

$29 \leq \dfrac{p^2 + 9}{2} < 45$

Give your answer as an inequality.

| LHS | RHS |
|---|---|
| $29 \leq \dfrac{p^2 + 9}{2}$ | $\dfrac{p^2 + 9}{2} < 45$ |
| $58 \leq p^2 + 9$ | $p^2 + 9 < 90$ |
| $49 \leq p^2$ | $p^2 < 81$ |
| $7 \leq p$ | $-9 < p < 9$ |
| $so \ p \geq 7$ | |

→ Solve each side as a separate inequality.

Sketching a graph might help.

$-9 < p \leq -7$

$7 \leq p < 9$     → Combine the two solutions.

# Iterations

**Iteration** means repeating a process. It can be used to find an approximate solution to more complex equations.

To find the iterative formula for a cubic equation, e.g. $x^3 + bx^2 - c = 0$, rearrange the equation as follows:

$$x_{n+1} = \sqrt[3]{c - (bx_n)^2}$$

Substitute the initial value of $x$ into the equation to find a solution. Then substitute the new value of $x$ into the equation until you find two solutions that are the same to the specified place (e.g. 2 d.p.) The solution lies between these two values.

## Worked example

**(a)** Show that the equation $x^3 + x^2 - 3 = 0$ has a solution between $x = 1$ and $x = 2$

$1^3 + 1^2 - 3 = -1$      Substitute $x$ into the equation.

$2^3 + 2^2 - 3 = 9$      Substitute $x = 2$ into the equation.

The answers have different signs, so there must be a solution between $x = 1$ and $x = 2$      Give a reason.

**(b)** Starting with $x_0 = 1$, use the iterative formula

$x_{n+1} = \sqrt[3]{3 - (x_n)^2}$ to work out $x^3 + x^2 - 3 = 0$ to 2 significant figures.

$x_1 = \sqrt[3]{3 - 1^2} = 1.2599\ldots$      Substitute $x_0 = 1$ into the iterative formula.

$x_2 = \sqrt[3]{3 - 1.2599\ldots^2} = 1.1220\ldots$

$x_3 = \sqrt[3]{3 - 1.1220\ldots^2} = 1.2030\ldots$      Substitute your answer into the iterative formula, and repeat until you get two answers that round to the same number to 2 s.f.

$x_4 = \sqrt[3]{3 - 1.2030^2} = 1.1579\ldots$

$x_4$ and $x_4$ are both 1.2 to 2 s.f., so the estimate is $x \approx 1.2$ (2 s.f.)

> **REVISION TIP**
>
> You need to use your calculator efficiently for iteration questions. Start by typing in the value of $x_0$. For example, for part **(b)**, first press [ 1 ]. Then press [ = ].
>
> Now enter the formula like this:
>
>   [ 3 ]  [ ANS ] [ $x^2$ ]

---

**Key terms**   Make sure you can write a definition for these key terms.

iteration   line of symmetry   quadratic graph
roots   sketch   turning point

## 10 Quadratic graphs, iterations, solving quadratic inequalities

Learn the answers to the questions below, then cover the answers column with a piece of paper and write as many as you can. Check and repeat.

| | Questions | Answers |
|---|---|---|
| 1 | What is the general equation of a quadratic graph? | $y = ax^2 + bx + c$ |
| 2 | What are the roots of a quadratic equation? | The $x$-intercept(s). |
| 3 | What is a turning point on a graph? | A maximum or minimum point. |
| 4 | Describe a quadratic graph with $a$ negative $x^2$ term. | ∩-shaped curve. |
| 5 | When using an iterative formula, what does a change of sign between two $x$-values tell you? | That the equation has a root between those two $x$-values. |
| 6 | What should be labelled in a sketch of a quadratic graph? | The $y$-intercept, the $x$-intercept(s), and the turning point. |
| 7 | Where is the line of symmetry on a quadratic graph? | On the vertical line that passes through the turning point. |
| 8 | If the quadratic graph has one $x$-intercept, what does this mean? | There is one real root. |
| 9 | If a quadratic graph has no $x$-intercepts, what does this mean? | There are no real roots. |
| 10 | How do you solve a quadratic inequality? | Factorise or complete the square the quadratic expression to find the turning points. Sketch the curve. Determine the region of the graph that satisfies the inequalities. |

*Put paper here*

## Previous questions

Now go back and use these questions to check your knowledge of previous topics.

| | Questions | Answers |
|---|---|---|
| 1 | When you are asked to estimate the value of a calculation, what should you round each number in the calculation to? | 1 significant figure. |
| 2 | On a distance–time graph, what does a horizontal section represent? | No movement; the speed is 0. |
| 3 | How do you divide by a fraction? | Multiply by its reciprocal. |
| 4 | What is the general equation of a straight line? | $y = mx + c$ |
| 5 | What is the general form of a quadratic expression? | $ax^2 + bx + c$ |

*Put paper here*

# Practice

**10**

## Exam-style questions

 **10.1** Solve each of these inequalities. Represent your solutions on the number lines.

(a) $x^2 - 4 \leq -3$

**[3 marks]**

(b) $7x^2 \geq 28$

**[3 marks]**

 **10.2** Solve each of these inequalities. Write your answers using set notation.

(a) $x^2 - 8x + 15 \leq 0$ **[4 marks]**

(b) $3x^2 - x - 4 > 0$ **[4 marks]**

(c) $x^2 - 30 < -7x$ **[4 marks]**

 **10.3** $n$ is an integer such that $2n + 7 \geq 3$ and $\dfrac{12 - n}{n^2} > 1$

Work out all the possible values of $n$. **[5 marks]**

 **10.4** Solve $14 < \dfrac{w^2 - 7}{3} < 31$ **[5 marks]**

 **10.5** (a) Show that the equation $x^4 - 12x = 0$ has a solution between 2 and 3 **[2 marks]**

(b) Starting with $x_0 = 2$, use the iterative formula $x_{n+1} = \sqrt[4]{12x_n}$ to work out an estimate for a solution of $x^4 - 12x = 0$ to 3 decimal places. **[3 marks]**

# Exam-style questions

**10.6** The graph with equation $y = x^3 + 5x^2 - 1$ is shown.

Dara wants to find the roots of the graph.

**(a)** Explain why Dara needs to solve the equation $x^3 + 5x^2 - 1 = 0$    **[1 mark]**

**(b)** Show that the equation $x^3 + 5x^2 - 1 = 0$ can be rearranged to

$$x = \frac{1 - 5x^2}{x^2}$$    **[2 marks]**

**(c)** Starting with $x_0 = -4$, use the iterative formula

$$x_{n+1} = \frac{1 - 5(x_n)^2}{(x_n)^2}$$

to work out a root of the graph to 2 decimal places.    **[3 marks]**

**(d)** Show that the equation $x^3 + 5x^2 - 1 = 0$ can be rearranged

to $x = \sqrt{\dfrac{1 - x^3}{5}}$

**(e)** Starting with $x_0 = 0$, use the iterative formula

$$x_{n+1} = \sqrt{\frac{1 - (x_n)^3}{5}}$$

to work out another root of the graph to 2 decimal places.    **[3 marks]**

**(f)** Starting with $x_0 = -1$, use the iterative formula

$$x_{n+1} = \frac{1 - (x_n)^3}{5x_n}$$

to work out the final root of the graph to 1 decimal place.    **[3 marks]**

**10.7** Given that $x^2 + 6x + 10 = (x + a)^2 + b$

    **(a)** Work out the value of $a$ and the value of $b$.       **[2 marks]**

    **(b)** Hence write down the coordinates of the turning point of the graph $y = x^2 + 6x + 10$       **[1 mark]**

    **(c)** Sketch the graph, clearly labelling the turning point and any intercepts.       **[2 marks]**

### Questions referring to previous content

 **10.8** Expand and simplify $(x - 1)(2 - x)(x + 4)$.       **[4 marks]**

 **10.9** Solve the equation $x^2 + 25 = 6x$ by rearranging and completing the square.       **[3 marks]**

# ⚙ Knowledge

## 11 Solving simultaneous equations

## Simultaneous equations

In **simultaneous equations** the variables have the same values. You need to combine the equations to be able to find both values.

For example, for this pair of simultaneous equations:

$2x + y = 5$ and $x - y = 1$

the values $x = 2$ and $y = 1$ are true for both equations:

$2 \times 2 + 1 = 5$ and $2 - 1 = 1$

## Solving simultaneous equations using graphs

Simultaneous equations can be solved using graphs.

The solution to a pair of simultaneous equations is the point at which their graphs **intersect**.

---

### Worked example

Use a **graphical method to** find the solutions of the simultaneous equations:

$y = 2x + 7$ and $y = -x + 4$

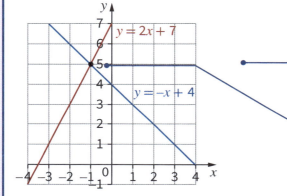

Draw both lines on the same axes.

$y = 2x + 7$ has gradient 2 and $y$-intercept 7

$y = -x + 4$ has gradient $-1$ and $y$-intercept 4

Find the point where the lines intersect. This is the solution to both equations.

The solution is $x = -1$, $y = 5$

**REVISION TIP**

You might need to rearrange the equations before you can plot them on the axes.

**LINK**

To remind yourself about drawing straight-line graphs, look back at Chapter 6.

Check that $x = -1$, $y = 5$ is a solution for both equations

$y = 2x + 7$        $y = -x + 4$

$5 = 2x - 1 + 7$     $5 = --1 + 4$

$5 = -2 + 7$ ✓      $5 = 1 + 4$ ✓

# Solving simultaneous equations using graphs

**Worked example**

1. Use a graphical method to estimate, to 1 decimal place, the solutions to the simultaneous equations $y = x^2 - 2x$ and $y = x + 1$

| $x$ | −1 | 0 | 1 | 2 | 3 |
|-----|----|----|----|----|----|
| $y$ | 3 | 0 | −1 | 0 | 3 |

Draw a table of values for the quadratic graph $y = x^2 - 2x$

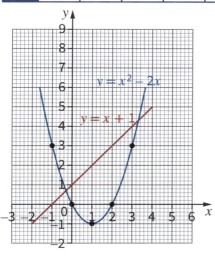

Plot the two graphs.

$y = x + 1$ is a straight line with gradient 1 and $y$-intercept 1.

The solutions are approximately
$x = -0.3$, $y = 0.7$ and $x = 3.3$, $y = 4.3$

Find the coordinates of the points where the two lines intersect. These are the solutions.

2. Using the graph of $y = x^2 - 2x$, estimate the solution to the equation $y = x^2 - 2x + 2$

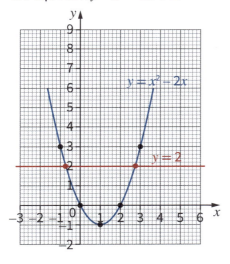

The graph of $y = x^2 - 2x$ has been given and solved.

Add 2 to both sides to obtain the new equation:

$y = x^2 - 2x + 2$

Since you are given $x^2 - 2x$, add 2 to both sides giving $2 = x^2 - 2x$

Then draw the line on the graph $y = 2$.

Approximately $x = -0.8$ and $x = 2.8$

Use the intersection of the graphs to estimate the solutions.

# ⚙ Knowledge

## 11 Solving simultaneous equations

## The substitution method

Sometimes, it is easier to **substitute** one variable into another; for example, if one example has a variable as the subject.

**Worked example**

Solve the simultaneous equations.

$y = x + 2$  ①          $2x + 3y = 11$ ②

Start with the more complicated equation.

$2x + 3y = 11$

Using ①, substitute $x + 2$ in place of $y$ in equation ②.

$2x + 3(x + 2) = 11$

Multiply out the bracket.

$2x + 3x + 6 = 11$

Collect like terms.

$5x + 6 = 11$

Subtract 6 from both sides.

$5x = 5$

Divide by 5 to find the value of $x$

$x = 1$

$y = x + 2$

$y = 1 + 2$

$y = 3$

Substitute $x = 1$ into equation ① to find $y$.

> **REVISION TIP** 📝
>
> You can check your answers by substituting $x = 1$, $y = 3$ into equation ②:
>
> $2x + 3y = 2 \times 1 + 3 \times 3 = 2 + 9 = 11$ which is correct.

## The elimination method

There are two **algebraic methods** to solve simultaneous equations. This means solving them using algebra and not a graph.

In the **elimination method** you add or subtract the two equations to get rid of either $x$ or $y$.

**Worked example**

Solve the simultaneous equations
$2x + 3y = 25$ and $6x - 2y = -2$

$2x + 3y = 25$ ①

$6x - 2y = -2$ ②

Number the equations ① and ②.

$(2x + 3y = 25) \times 3 = 6x + 9y = 75$ ③

We need the $x$ or $y$ coefficient to be the same. Here, $6x$ is a multiple of $2x$, so multiply equation ① by 3. Label this equation ③.

$6x + 9y = 75$ ③

$-(6x - 2y = -2)$ ②

Both equations now have $6x$, so we can eliminate $x$ by subtracting equation ② from ③.

$0x + 11y = 77$

$y = 77$

Be careful subtracting negatives.

$2x + 3(7) = 25$

$2x + 21 = 25$

Substitute $y = 7$ into equation ① to find $x$.

$2x = 4$

$x = 2$

## The elimination method

**Worked example**

Check:

$2(2) + 3(7) = 25$    $6(2) - 2(7) = -2$

$\quad 4 + 21 = 25$    $\quad 12 - 14 = -2$

$\quad\quad 25 = 25$ ✔    $\quad\quad -2 = -2$ ✔

**REVISION TIP**

If the signs are the same: **Subtract**

$$5x \;\textcircled{+ 2y} = 16 \quad ①$$
$$\ominus (3x \;\textcircled{+ 2y} = 12) \quad ②$$
$$\overline{\quad 2x + 0y = 4 \quad}$$

If the signs are different: **Add**

$$5x \;\textcircled{+ 2y} = 16 \quad ①$$
$$\oplus (3x \;\textcircled{- 2y} = 12) \quad ②$$
$$\overline{\quad 8x + 0y = 28 \quad}$$

## Word-based problems

Sometimes you will be given a **word-based problem** from which you need to form two simultaneous equations.

**Worked example**

The cost of three adult tickets and two child tickets to a theme park is £155.

The cost of two adult tickets and five child tickets is £195.

**1.** Write a pair of simultaneous equations to describe this situation.

> Decide what the two variables represent and give them each a letter.

$a$ = cost of adult ticket

$c$ = cost of child ticket.

> Use the information in the question to form two equations using your variables. Number them ① and ② to help with the next part of the question.

$3a + 2c = 155$ ① and $2a + 5c = 195$ ②

**2.** Solve your equations to find the cost of an adult ticket and the cost of a child ticket.

> Multiply ① by 2: $6a + 4c = 310$ ③
>
> Multiply ② by 3: $6a + 15c = 585$ ④
>
> Both of the $6a$ terms are positive, so subtract equation 3 from equation 4.

④ − ③: $11c = 275$

$\quad\quad\quad c = 25$

> Substitute into equation ①: $3a + 50 = 155$

$3a = 105$, so $a = 35$

The cost of an adult ticket is £35 and the cost of a child ticket is £25.

**Key terms** **Make sure you can write a definition for these key terms.**

algebraic method    elimination method    graphical method
intersect    simultaneous equations    substitute
word-based problem

## 11 Solving simultaneous equations

Learn the answers to the questions below, then cover the answers column with a piece of paper and write as many as you can. Check and repeat.

| | Questions | | Answers |
|---|---|---|---|
| 1 | What are simultaneous equations? | | A pair of equations with two variables that you solve at the same time. |
| 2 | What are the two algebraic methods you can use to solve simultaneous equations? | | Elimination and substitution. |
| 3 | What does the point of intersection represent on two graphs? | | The solution to the pair of simultaneous equations. |
| 4 | To form simultaneous equations from word-based problems, what is the first step? | | Decide what the two variables represent. |
| 5 | How can you check your solutions to a pair of simultaneous equations? | | Substitute your values back into the equations. |
| 6 | How would you eliminate $x$ from the equations $3x + 5y = 1$ ① $3x - 2y = 8$ ② | | Subtract equation ② from equation ①. |
| 7 | How would you eliminate $y$ from the equations $2x - y = 2$ ① $7x + y = 43$ ② | | Add the two equations together. |
| 8 | What would be your first step if you were asked to solve these simultaneous equations by elimination? $2x + 3y = 7$ ① $x - 2y = 4$ ② | | Multiply one or both equations so that either $x$ or $y$ have the same coefficient in both equations. |
| 9 | If you form and solve simultaneous equations from a word problem, what should your final step be? | | Use your solutions to answer the original question in context. |

*Put paper here*

### Previous questions

Now go back and use these questions to check your knowledge of previous topics.

| | Questions | | Answers |
|---|---|---|---|
| 1 | What does the $a \geq b$? | | $a$ is greater than or equal to $b$. |
| 2 | What is an identity? | | An equation that is true for any value of the variable. |
| 3 | When a number is written in standard form $(a \times 10^n)$ what values can the index number, $n$, take? | | Any positive or negative integer. |
| 4 | What does it mean to expand a single bracket? | | To multiply each term inside the bracket by the term in front of the bracket. |

*Put paper here*

## Exam-style questions

**11.1**  Here are the graphs of $y = 2x - 3$ and $x + y = 4$

Use the graphs to estimate the solution to these simultaneous equations.

$y = 2x - 3$

$x + y = 4$

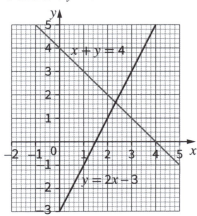

**EXAM TIP**

The solution of two simultaneous equations is where the lines intersect.

**[2 marks]**

**11.2**  Here is the graph of $x + 2y = 6$

By drawing a suitable line on the graph, find the solution to the simultaneous equations

$x + 2y = 6$

$y = x$

**[2 marks]**

**11.3**  Solve each pair of simultaneous equations.

| | | | |
|---|---|---|---|
| **(a)** | $x + y = 14$ | $x - y = 8$ | **[3 marks]** |
| **(b)** | $2x - 2y = 4$ | $2x + 3y = 14$ | **[3 marks]** |
| **(c)** | $4x + 5y = 37$ | $2x + y = 11$ | **[3 marks]** |
| **(d)** | $3x - 2y = 2$ | $12x - 4y = 10$ | **[3 marks]** |

**11.4**  Solve each pair of simultaneous equations.

| | | | |
|---|---|---|---|
| **(a)** | $2x + 5y = 11$ | $3x - 2y = -12$ | **[3 marks]** |
| **(b)** | $2x - 7y = 12$ | $5x - y = -3$ | **[3 marks]** |
| **(c)** | $3x + 8y = 12$ | $2x + 12y = 13$ | **[3 marks]** |
| **(d)** | $6x - 4y = 9$ | $5x + 3y = -2$ | **[3 marks]** |

**EXAM TIP**

You might need to multiply both equations before you can eliminate $x$ or $y$.

# Exam-style questions

**11.5** Two families go to the cinema.

The first family buys one adult ticket and three child tickets and pays £39

The second family buys two adult tickets and four child tickets and pays £62

    **(a)** Write a pair of simultaneous equations to describe this situation. **[2 marks]**

    **(b)** Work out the cost of an adult ticket and the cost of a child ticket. **[3 marks]**

**11.6** The mass of 20 apples and 30 satsumas is 4050 g.

The mass of 12 apples and 15 satsumas is 2205 g.

Work out the mass of one apple and the mass of one satsuma. **[4 marks]**

**EXAM TIP**

Start by writing a pair of simultaneous equations to represent the information.

**11.7** $2^x \times 2^y = 64$ and $2^x \div 2^y = 4$

Work out the values of $x$ and $y$. **[4 marks]**

**EXAM TIP**

Use the laws of indices to write two linear simultaneous equations.

**11.8** Use a graphical method to work out the solutions to the simultaneous equations

$y = (x - 2)^2$

$y = x$

**[4 marks]**

**11.9** Use an algebraic method to solve these pairs of simultaneous equations.

    **(a)** $x + y = 3$         $y = x^2 + 3x - 2$ **[4 marks]**

    **(b)** $2x^2 + y^2 = 9$     $2y + x = 0$ **[4 marks]**

**EXAM TIP**

Be careful when squaring. Look at the coefficients too.

**11.10** The graph of $y = x^2 - 3x - 18$ is shown.

**(a)** Use a graphical method to work out the solutions to the simultaneous equations

$y = x^2 - 3x - 18$

$y = 2x - 12$ **[3 marks]**

**(b)** Use the graph to estimate the solutions to the equation
$x^2 - 3x - 14 = 0$ **[3 marks]**

**Questions referring to previous content**

**11.11** Solve the inequality for $x$ and show your solution on the number line.

$11 + 2x \geq 5x - 1$

**[2 marks]**

**11.12** Solve $6x^2 - 7x + 20 > 0$

Give your answer in set notation. **[4 marks]**

# ⚙ Knowledge

## 12 Sequences

## Sequences: an overview

A **sequence** is an ordered set of numbers or patterns that follow the same rule.

The numbers in a sequence are called **terms**.
first number = first term
second number = second term
$n$th number = **$n$th term**

An **arithmetic sprogression** (or linear sequence) is one in which the next term can be found by adding or subtracting the same number each time.

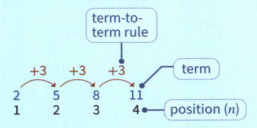

**Term-to-term rule**

The rule that gets you from one term to the next. This is the same for every term.

**Position-to-term rule**

A rule that finds the term based on where it is in the sequence. The position to term rule is always written in terms of $n$. This is called the $n$th term. In the example above, it is $3n - 1$.

## The term-to-term rule

**Worked example**

A sequence of patterns made from sticks are shown.

Pattern 1     Pattern 2     Pattern 3

1. Write down the term-to-term rule for this sequence.

   Make a table and count the number of matches for each pattern.

   | $n$ | 1 | 2 | 3 |
   |---|---|---|---|
   | sticks | 3 | 5 | 7 |

   (+2    +2)

   Two more sticks are added each time. So this is an arithmetic progression.

   The term-to-term rule is 'add 2'.

2. How many sticks will there be in Pattern 6?

   Continue adding 2.

   | $n$ | 1 | 2 | 3 | 4 | 5 | 6 |
   |---|---|---|---|---|---|---|
   | sticks | 3 | 5 | 7 | 9 | 11 | 13 |

   (+2  +2  +2  +2  +2)

   There are 13 sticks in Pattern 6.

## The position-to-term rule

**Worked example**

The $n$th term of a sequence is $11 + 3n$.

1. Work out the seventh term of the sequence.

   seventh term
   $= 11 + 3 \times 7 = 32$ ● — Substitute $n = 7$

   **REVISION TIP**

   The question gives you an expression for the $n$th term, so you know that $11 + 3n$ is the position-to-term rule.

2. Is 81 a term of this sequence?

   Show how you get your answer.

   $11 + 3n = 81$

   $\dfrac{3n}{3} = \dfrac{70}{3}$

   $n = 23.\dot{3}$

   70 is not a multiple of 3, so the solution is not an integer. Hence 81 is **not** in the sequence.

# Finding the *n*th term

The ***n*th term** is the position-to-term rule. You can use it to find any term in the sequence. *n* is the term number. To find the *n*th term, follow the steps in the worked example.

**Worked example**

Work out an expression for the *n*th term of the sequence: 5, 9, 13, 17

5   9   13   17
+4   +4   +4

> Work out the common difference between terms.

Common difference = 4

$4n$:   4   8   12   16

> Write out the multiples of the common difference.

↓+1 ↓+1 ↓+1 ↓+1
5   9   13   17

> Work out what to add or subtract to get the original sequence.

$n$th term $= 4n + 1$

> Write out the *n*th term rule.

$(4 \times 2) + 1 = 9$

> Check your answer by substituting in $n = 2$.

# Special sequences

Special sequences do not have a common difference between their terms. They are not arithmetic sequences.

**Fibonacci sequences**

Any sequence where you add the previous two terms to get the next term is a Fibonacci-*type* sequence. The 'original' **Fibonacci sequence** was:

1, 1, 2, 3, 5, 8, 13, 21 ...

**Worked example**

The first three terms of a Fibonacci-type sequence are $k, k + 6, 2k + 6$
Work out the fourth term of the sequence.

$= (k + 6) + (2k + 6)$
$= 3k + 12$

> 4th term = 2nd term + 3rd term

The first five **square numbers**:

 $2^2 = 4$

 $3^2 = 9$

 $4^2 = 16$

 $5^2 = 25$

 $6^2 = 36$

The first three **cube numbers**:

 $1^3 = 1$

 $2^3 = 8$

 $3^3 = 27$

The first four **triangular numbers**:

1

$1 + 2 = 3$

$1 + 2 + 3 = 6$

$1 + 2 + 3 + 4 = 10$

**REVISION TIP**

To find the next triangular number, just add on the next positive integer.

## 12 Sequences

### Quadratic sequences

In a **quadratic sequence**, the position-to-term rule includes a squared term.

---

**Worked example**

1. Here are the first five terms of a sequence.

   1, 5, 12, 22, 35

   Work out the next term of this sequence.

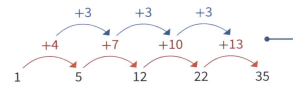

> Calculate:
> - the difference between each term
> - the difference between the first differences.

   next difference: $13 + 3 = 16$
   next term: $35 + 16 = 51$

> Find the difference to the next term.
> Use this to find the next term.

2. The $n$th term of a different sequence is $n^2 - 2n$.

   Work out the 3rd term of the sequence.

   $n^2 - 2n$

   3rd term $= 3^2 - (2 \times 3)$

   $= 3$

> To find the 3rd term of the sequence, substitute $n = 3$

---

**Worked example**

The first four terms of a quadratic sequence are: 4, 5, 12, 25, …

Work out an expression for the $n$th term.

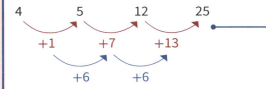

> Find the second difference.

Sequence involves $3n^2$.

> Halve the second difference and multiply it by $n^2$.

| $n$ | 1 | 2 | 3 | 4 |
|---|---|---|---|---|
| | 4 | 5 | 12 | 25 |
| $3n^2$ | 3 | 12 | 27 | 48 |
| | 1 | −7 | −15 | −23 |

> Subtract $3n^2$ from each term of the original sequence.

| $n$ | 1 | 2 | 3 | 4 |
|---|---|---|---|---|
| | 1 | −7 | −15 | −23 |

> Find the $n$th term of the new linear sequence:
> 1, −7, −15, −23

difference between terms is −8

$n$th term is $-8n + 9$

The $n$th term of the quadratic sequence is
$3n^2 - 8n + 9$

> Put this together with the quadratic term you found earlier.

**REVISION TIP**

Check the $n$th term by substituting in a value for $n$.

---

## Geometric progressions

The next term in a **geometric progression** is found by multiplying or dividing by the same number each time. For example, the term-to-term rule for this sequence is '×3'

2    6    18    54
×3  ×3  ×3

**Worked example**

1. A sequence has $n$th term rule $8^n$

   (a) Work out the 3rd term of the sequence.

   $8^3 = 512$ ← Substitute 4 in place of $n$

   (b) What is the term-to-term rule?

   Multiply by 8

   To find the term-to-term rule, divide a term by the preceding term. Check it holds true for at least 2 instances.

2. A sequence has $n$th term rule $(\sqrt{5})^n$

   (a) Work out the 4th term of the sequence.

   $(\sqrt{5})^4$ ← Substitute 4 in place of $n$, then simplify the surd.
   $= \left(5^{\frac{1}{2}}\right)^4$
   $= 5^2$
   $= 25$

   (b) What is the term-to-term rule?

   The term-to-term rule 'multiply by $\sqrt{5}$'

   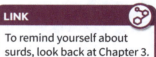

   **LINK**
   To remind yourself about surds, look back at Chapter 3.

**Worked example**

The first three terms of a sequence are $(\sqrt{x} - 2)$, 1, $(\sqrt{x} + 2)$

(a) Find the term-to-term rule.

$$\frac{1}{(\sqrt{x} - 2)} = \frac{(\sqrt{x} + 2)}{1}$$

Use the fact that the second term divided by the first term is equal to the third term divided by the second term to write an algebraic equation.

$$(\sqrt{x} - 2)(\sqrt{x} + 2) = 1$$
$$x - 2\sqrt{x} + 2\sqrt{x} - 4 = 1$$

Multiply both sides by $(\sqrt{x} - 2)$ then, expand the brackets and solve it to find the value of $x$.

$$x - 4 = 1$$
$$x = 5$$

Find the term-to-term rule by dividing the third term by the second term and substitute the value for $x$.

(b) The term-to-term rule is 'multiply by $(\sqrt{5} + 2)$'.
Show that the 4th term is $4\sqrt{5} + 9$

4th term $(\sqrt{5} + 2)(\sqrt{5} + 2)$

Substitute '$x = 5$' into third term then multiply it by the term-to-term rule found in part (a).

$$= 5 + 2\sqrt{5} + 2\sqrt{5} + 4$$
$$= 4\sqrt{5} + 9$$

Expand and simplify.

**Key terms** — Make sure you can write a definition for these key terms.

arithmetic progression   cube numbers   geometric progression
fibonacci-type sequence   nth term   position-to-term rule
quadratic sequence   term   term-to-term rule
triangular numbers   linear sequence   square numbers

## 12 Sequences

Learn the answers to the questions below, then cover the answers column with a piece of paper and write as many as you can. Check and repeat.

| Questions | Answers |
|---|---|
| **1** What is an arithmetic progression? | A sequence that increases or decreases by the same amount each time. |
| **2** How do you get to the next term of a geometric progression? | Multiply or divide by the same number each time. |
| **3** What is the term-to-term rule? | The rule that gets you from one term to the next. This is the same for every term. |
| **4** What can you use the $n$th term of a sequence for? | To find any term in the sequence. |
| **5** Is 14 in the sequence $3n - 1$? | Let $14 = 3n - 1$<br>So $15 = 3n \Rightarrow n = 5$<br>Since $n = 5$ is an integer, 14 is in the sequence. |
| **6** How do you find the next term in a Fibonacci sequence? | Add the previous two terms to get the next one. |
| **7** Is 15 a triangular number? How do you know? | Yes, because $1 + 2 + 3 + 4 + 5 = 15$. |
| **8** What is a cube number? | A number multiplied by itself three times, such as $3 \times 3 \times 3 = 27$. |
| **8** What is the general form of a quadratic sequence? | $ax^2 + bx + c$ |
| **9** How do you find the common ratio of a geometric progression? | Divide any term by the previous term. |
| **10** What is a linear sequence? | A sequence that increases or decreases by the same amount. |
| **11** What are the first five terms of the square number sequence? | 1, 4, 9, 16, 25 |
| **12** True or false? Sequences always involve numbers. | False; you could be asked to draw a sequence using patterns. |

*Put paper here*

### Previous questions

Now go back and use these questions to check your knowledge of previous topics.

| Questions | Answers |
|---|---|
| **1** How do you represent an inequality on a number line where the value is not included? | With an open circle. |
| **2** How do you find 50% of a number? | Divide by 2 or multiply by 0.5. |
| **3** How do you write a ratio in its simplest form? | Divide by the highest common factor. |

*Put paper here*

## Exam-style questions

**12.1** The $n$th term of a sequence is $35 - 4n$.

    **(a)** Work out the value of the first negative term in the sequence. **[2 marks]**

    **(b)** Show that $-100$ is not a number in this sequence. **[2 marks]**

**12.2** The $n$th term of a sequence is $8n + 3$.

    **(a)** Which term in the sequence is 51? **[2 marks]**

    **(b)** Show that 64 is not a number in the sequence. **[2 marks]**

    **(c)** Work out the value of the first term in the sequence that
is greater than 100. **[3 marks]**

**EXAM TIP**
Write and solve an inequality.

The $n$th term of a sequence is $n^2 - 30$.

    **(a)** Find the 4th term in the sequence. **[2 marks]**

    **(b)** Is the number 114 in this sequence?

    Explain your answer. **[2 marks]**

**12.3** A Fibonacci-type sequence starts $m, n, m + n, \dots$

    **(a)** Work out

        **(i)** the 4th term in the sequence **[1 mark]**

        **(ii)** the 7th term in the sequence. **[2 marks]**

    **(b)** The 1st term of the sequence is 3.

    The difference between the 1st and 3rd terms is 5.

    Work out the 8th term of the sequence. **[2 marks]**

**12.4** A geometric progression starts $\frac{1}{2}, \frac{1}{6}, \frac{1}{18}, \dots$

    **(a)** Write the term-to-term rule.

    **(b)** Write an expression, in terms of $n$, for the $n$th term of
the sequence. **[15 marks]**

**12.5** An arithmetic progression has 1st term 5 and 3rd term 11.

Find an expression, in terms of $n$, for the $n$th term of the sequence? **[3 marks]**

**EXAM TIP**
Work out some terms in the sequence first.

**12.6** A geometric progression starts 12, 9, 6, 3, ...

Work out the sum of the 50th and 60th terms in this sequence. **[3 marks]**

**12.7** Here is a number sequence: $3, \frac{5}{2}, \frac{7}{3}, \frac{9}{4}, \frac{11}{5}, \ldots$

**(a)** Write the next term in the sequence. **[1 mark]**

**EXAM TIP**
Look at the numerators and denominators separately.

**(b)** Write an expression, in terms of $n$, for the $n$th term of the sequence. **[3 marks]**

**(c)** Work out the product of the 6th and 9th terms of the sequence. **[2 marks]**

 **12.8** A quadratic progression starts 1, 12, 27, 46, ...

Write an expression, in terms of $n$, for the $n$th term of the sequence. **[6 marks]**

**12.9** A quadratic progression starts −1, −5, −11, −20, ...

Write an expression, in terms of $n$, for the $n$th term of the sequence. **[6 marks]**

 **12.10** A sequence has $n$th term $n^2 + 2n + 2$

Work out which term in the sequence is 50.

**EXAM TIP**
Write and solve an equation.

 **12.11** The $n$th term of sequence is given by $n^2 + 2n + 5$, where $b$ and $c$ are integers.

The 2nd term of the sequence is 13.

The 5th term of the sequence is 40.

Write an expression, in terms of $n$, for the $n$th term of the sequence. **[5 marks]**

 **12.12** The $n$th term of sequence is given by $an^2 + b$, where $a$ and $b$ are integers.

The 4th term of the sequence is 42.

The 9th term of the sequence is 237.

Find the 15th term in the sequence. **[5 marks]**

 **12.13** The first three terms of a geometric progression are

$9 + \sqrt{y}, \ 4, \ 9 - \sqrt{y}$

Work out the value of $y$.

You must show your working. **[3 marks]**

**12.14 (a)** A sequence has the $n$th term $\frac{n}{n + 3}$

Write the first three terms of the sequence. **[2 marks]**

**(b)** Hence, write an expression, in terms of $n$, for the $n$th term of the sequence

$\frac{3}{5}, \frac{4}{7}, \frac{5}{9}, \frac{6}{11}, \frac{7}{13}, \ldots$

**[2 marks]**

**12.15 (a)** A geometric progression has the $n$th term $\dfrac{\left(\sqrt{3}\right)^{n}}{3}$

Work out the first three terms. **[3 marks]**

**(b)** Write an expression, in terms of $n$, for the $n$th term of the sequence

$5\sqrt{2},\ 10,\ 10\sqrt{2},\ 20,\ 20\sqrt{2},\ \dots$ **[2 marks]**

**Questions referring to previous content**

 **12.16** Solve the simultaneous equations **[4 marks]**

$3y - 4x^2 = 18$ and $10x = y - 32$

 **12.17** Harriet buys three pineapples and six bananas for £17.10.

Eric buys four pineapples and 9 bananas for £24.05.

Work out the cost of one pineapple and one banana. **[5 marks]**

# ⚙ Knowledge

## 13 Cubic graphs, reciprocal graphs, exponential graphs, transformation of graphs

## Cubic and reciprocal graphs

A **cubic graph** has an $x^3$ term in its equation.

For example, $y = x^3 + x^2 + 4x + 7$

The general equation of a cubic graph is:

$y = ax^3 + bx^2 + cx + d$

where $a$, $b$, $c$, and $d$ are constants.

A **reciprocal graph** is of the form $y = \dfrac{a}{x}$,

where $a$ is a constant.

When $y = \dfrac{1}{x}$, the value of $x$ cannot be 0,

because $\dfrac{1}{0}$ is not a defined value.

The curve will get closer and closer to the axes but will never touch them.

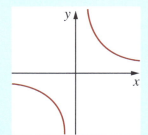

**REVISION TIP** ☑

An inverse proportion graph is a type of reciprocal graph. You will revise inverse proportion in Chapter 17.

### Worked example

Match each graph with a possible equation.

$y = 2x - 1$    $y = 2x^2 - 1$    $y = 2x^3 - 1$    $y = \dfrac{2}{x}$

Graph is a cubic
$y = 2x^3 - 1$

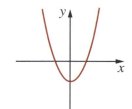

Graph is a quadratic
$y = 2x^2 - 1$

Graph is a straight line
$y = 2x - 1$

Graph is a reciprocal
$y = \dfrac{2}{x}$

### Worked example

1. Complete the table of values for
   $y = x^3 + 1$

   | $x$ | −2 | −1 | 0 | 1 | 2 |
   |---|---|---|---|---|---|
   | $y$ | −7 | 0 | 1 | 2 | 9 |

2. Draw the graph of $y = x^3 + 1$

   It is cubic, so we expect a ∿ shape.

**REVISION TIP**

If it's a cubic graph, look for $x^3$ in the equation.

If it's a quadratic graph, look for $x^2$ in the equation.

## Cubic and reciprocal graphs

**Worked example**

The time, $t$ hours, taken to complete a journey at speed $v$ km/h is given by the equation $t = \dfrac{75}{v}$.

1. Find the missing values for $t$ and $v$.

| $v$ (km/h) | 1 | 3 | 5 | $\dfrac{75}{7.5} = 10$ |
|---|---|---|---|---|
| $t$ (hours) | $\dfrac{75}{1} = 75$ | 25 | $\dfrac{75}{5} = 15$ | 7.5 |

2. Plot a graph of $t$ against $v$.

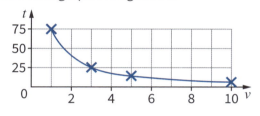

**REVISION TIP**

Start by drawing a sketch of the type of graph you expect, so that you can check it once you have plotted it.

## Exponential graphs

Functions where the variable is in the power are called **exponential** functions.

An exponential function is in the form $y = k^x$, where $k$ is a positive constant.

The graph of $y = k^x$ has $y$-intercept 1 (since $k^0 = 1$ for any value of $k$).

The graph of $y = Ak^x$ will have $y$-intercept $A$.

$y = 3^x$

$k > 1$ indicates exponential growth

$y = \left(\dfrac{1}{2}\right)^x$

$k < 1$ indicates exponential decay

**Worked example**

The number, $N$, of bacteria in a Petri dish at the start of an experiment is 300.
The number of bacteria in the dish after $t$ hours is given by $N = A \times 1.05^t$, where $A$ is a constant.

**(a)** Work out the value of $A$.

$A \times 1.05^0 = 300$ •————

$A = 300$ since $1.05^0 = 1$

Initially, $t = 0$ and $N = 300$

Substitute into the equation.

**(b)** How many bacteria will be in the dish after seven hours?

number of bacteria $= 300 \times 1.05^7 = 422$ •————

Substitute into the equation.

**(c)** Sketch the graph of $N$ against $t$.

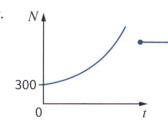

Sketch an exponential graph, with $y$-intercept of 300. Make sure you label the axes.

# Knowledge

## 13 Cubic graphs, reciprocal graphs, exponential graphs, transformation of graphs

### Graph transformations

The graph of $y = f(x)$, can be **translated** in the following ways:

- $y = f(x + a)$ is a translation of $a$ units left

- $y = f(x - a)$ is a translation of $a$ units right

- $y = f(x) + a$ is a translation of $a$ units up

- $y = f(x) - a$ is a translation of $a$ units down

**REVISION TIP**

For $y = f(x)$

- the $x$ is inside the brackets, so $f(x + a)$ will change the $x$-coordinate

- the $y$ is outside the brackets, so $f(x) + a$ will change the $y$-coordinate.

**WATCH OUT**

Pay attention to the positive and negative signs, and the impact they have on whether the graph moves to the left, right, up or down.

**LINK**

You will revise **transformations** and **translations** in Chapter 23.

---

**Worked example**

Here is the graph of $y = f(x)$

Sketch the graph of $y = f(x - 2)$

The graph is translated 2 units to the right. Add 2 to each of the $x$-intercepts.

---

**Worked example**

Here is the graph of $y = f(x)$

Sketch the graph of $y = f(x) + 2$

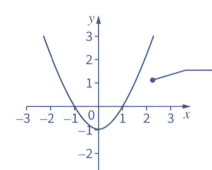

The graph is translated 2 units up. Add 2 onto the $y$-intercept.

# Graph transformations

If you have the graph of $y = f(x)$, then:

- $y = -f(x)$ is a reflection in the $x$-axis
- $y = f(-x)$ is a reflection in the $y$-axis

> **REVISION TIP**
>
> Remember this…
>
> The $x$ is inside the brackets, so $f(-x)$ will change the $x$-coordinate.
>
> The $y$ is outside the brackets, so $-f(x)$ will change the $y$-coordinate.

### Worked example

Here is the graph of $y = f(x)$

Sketch the graph of $y = -f(x)$

> The graph is reflected in the $x$-axis.
> The $x$-intercepts stay the same.
> Only the $y$-coordinates are affected.

### Worked example

Here is the graph of $y = f(x)$

Sketch the graph of $y = f(-x)$

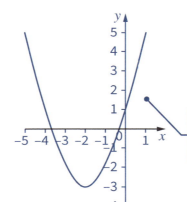

> The graph is reflected in the $y$-axis.
> The $y$-intercepts stay the same.
> Only the $x$-coordinates are affected.

### Worked example

The graph of the curve $C$, with equation $y = f(x)$, is transformed to give the graph of the curve $D$, with equation $y = -f(x + 5)$

The point on $C$ with coordinates $(3, 5)$ is mapped to the point $R$ on curve $D$.

Find the coordinates of point $R$.

$x$-coordinate: $3 - 5 = -2$

$y$-coordinate: $5x - 1 = -5$

The coordinates of $R$ are $(-2, -5)$

> The 5 inside the brackets tells you to subtract 5 from the $x$-coordinate.
>
> The negative sign outside the brackets tells you to multiply the $y$-coordinate by $-1$

**Key terms**

Make sure you can write a definition for these key terms.

cubic graph    exponential    reciprocal graph
transformation    translation

## 13 Cubic graphs, reciprocal graphs, exponential graphs, transformation of graphs

Learn the answers to the questions below, then cover the answers column with a piece of paper and write as many as you can. Check and repeat.

| | Questions | Answers |
|---|---|---|
| 1 | What is the general equation of a cubic graph? | $y = ax^3 + bx^2 + cx + d$ |
| 2 | What are the roots of a cubic equation? | The $x$-intercepts. |
| 3 | How do you find the average gradient of a curve between two points? | Draw a chord and find the gradient. |
| 4 | How do you find the gradient of a curve at a point? | Draw a tangent and find the gradient. |
| 5 | What is an exponential function? | A function of the form $\frac{1}{a}^x$. |
| 6 | What is the $y$-intercept of $y = k^x$? | $y = 1$, as $k^0 = 1$ |
| 7 | What transformation of f($x$) does $y = $ f($x + a$) represent? | Translation of $a$ units left. |
| 8 | What transformation of f($x$) does $y = $ f($x$) $+ a$ represent? | Translation of $a$ units up. |
| 9 | What transformation of f($x$) does $y = -$f($x$) represent? | Reflection in the $x$-axis. |
| 10 | What transformation of f($x$) does $y = $ f($-x$) represent? | Reflection in the $y$-axis. |

*Put paper here*

## Previous questions

Now go back and use these questions to check your knowledge of previous topics.

| | Questions | Answers |
|---|---|---|
| 1 | When you see a real-life graph, what should you do first? | Find out what each unit on each scale represents. |
| 2 | If there are 10 squares between 0 and 20, what does each square represent? | 2 units. |
| 3 | On a distance–time graph, what does the gradient represent? | Speed. |
| 4 | On a velocity–time graph, what does the gradient represent? | Acceleration. |
| 5 | On a distance–time graph, what does a horizontal section represent? | No movement. |

*Put paper here*

## Exam-style questions

**13.1** The sketch shows the graph of $y = \dfrac{a}{bx}$ where $a > 0$ and $b > 0$

Work out the values of $a$ and $b$. **[3 marks]**

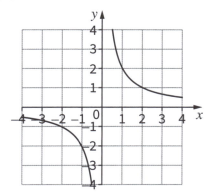

**13.2** The sketch shows the graph of $y = ab^x$, where $a > 0$ and $b > 0$.

The graph passes through the points $(1, 10)$ and $(-1, 0.4)$.

Work out the values of $a$ and $b$.

**[3 marks]**

**13.3** Here are four graphs.

Graph A     Graph B     Graph C

 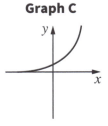

The graphs represent four different types of function.

Match each description of the function in the table to the letter of the graph.

| Description of function | Graph |
|---|---|
| Exponential | |
| Cubic | |
| Reciprocal | |

**[2 marks]**

**13.4** Sketch the curve with equation $y = 10^x$

Give the coordinates of any points of intersection with the axes.

**[3 marks]**

**13.5** The graph of the curve $C$ with equation $y = f(x)$ is transformed to give the graph of the curve $T$ with equation $2f(x - 2)$.

The point on $C$ with coordinates $(3, 5)$ is mapped to point $P$ on $T$.

Work out the coordinates of $P$. **[2 marks]**

**13.6** Here is the graph of $y = f(x)$

On the same set of axes, sketch the graph of $y = -f(x)$.

**[2 marks]**

**13.7** Graph $A$ has been transformed to give graph $B$.

The equation of graph $A$ is $y = f(x)$

Write down the equation of graph $B$.

**[1 mark]**

**13.8** Sketch the graph of $y = 0.5x$ for $-2 \leq x \leq 3$

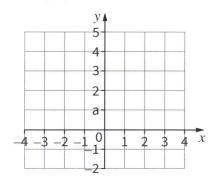

**[3 marks]**

**13.9** Here is the graph with equation $y = x^2 - 4x + 1$

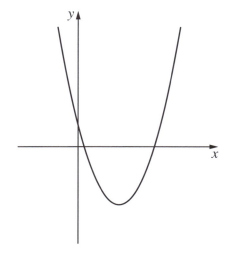

(a) Give the coordinates of the minimum point of the graph
with equation $y = x^2 - 4x + 1$ **[2 marks]**

(b) On the same axes, sketch the graph with equation
$y = (x + 1)^2 - 4(x + 1) + 1$ and give the coordinates of the
minimum point. **[1 mark]**

**Questions referring to previous content**

 **13.10** Solve algebraically the pair of simultaneous equations

$5x + y = 10 \qquad 10xy = -48$ **[5 marks]**

 **13.11** The first term of an arithmetic sequence is 4 and the fourth term is 25.
Find the common difference of the sequence and determine
if 61 is a term in the sequence. **[4 marks]**

## 14 Non-linear real-life graphs

## Gradients on non-linear real-life graphs

To calculate the average **gradient** between two points:

- draw a **chord** connecting the two points
- calculate the gradient of the chord.

To estimate the gradient at a point:

- draw a **tangent** to the curve at that point
- calculate the gradient of the tangent.

For a distance—time graph:

- the gradient of the chord gives average speed between the two points
- the gradient of the tangent gives an estimate of the speed at that point.

Gradient of tangent is estimate of speed at this point.

Gradient of chord is average speed between the two points.

**LINK**

You will revise tangents and chords in Chapter 19.

---

**Worked example**

The distance–time graph for a train on part of its journey is shown.

**(a)** Calculate the train's average speed in km/h during the first four minutes.

Draw a chord connecting the points on the curve where $x = 0$ and $x = 4$

Draw a tangent to the graph at $x = 10$ and a triangle to work out the gradient.

$$\text{gradient of chord} = \frac{\text{change in } y}{\text{change in } x}$$
$$= \frac{1.6}{4}$$
$$= 0.4$$

Find the gradient of the chord.

**WATCH OUT** ❗

Pay attention to the scale of the axes!

**(b)** Estimate the train's speed after 10 minutes in kilometres per hour.

$$\text{gradient of tangent} = \frac{1.6}{4} = 0.4$$
$$\text{speed} = 0.4\,\text{km/min}$$
$$= 24\,\text{km/h}$$

Calculate the gradient of the tangent to find the speed in km/min.

Convert the speed to km/h.

## Area under the curve on non-linear real-life graphs

For a velocity–time graph:

- the gradient of the tangent gives an estimate of the acceleration at that point
- the area under the graph gives the distance travelled up to that point.

Gradient of tangent is an estimate of the acceleration at this point.

Area under graph is the distance travelled up to this point.

## Area under the curve on non-linear real-life graphs

When estimating the area under a curve, the answer could be an over-estimate, or an under-estimate.

Your answer could either be
an under-estimate    or    an over-estimate

of the actual area.

### Worked example

Here is the velocity–time graph for an object in motion.

**(a)** Estimate the distance travelled by the object in the first 12 s.

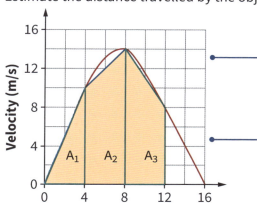

> Divide the area into three strips of equal width by drawing vertical lines at 4 s, 8 s, and 12 s.

> Draw a chord between the two outer points of each strip to approximate the curve.

$A_1 = \frac{1}{2} \times (10 + 0) \times 4 = 20$

> Calculate the area of each trapezium.

$A_2 = \frac{1}{2} \times (10 + 14) \times 4 = 48$

$A_3 = \frac{1}{2} \times (14 + 8) \times 4 = 44$

$20 + 48 + 44 = 112$, so the object travelled $112\,m$.

> Add the areas together to find the distance travelled.

**(b)** Is your answer to part **(a)** an overestimate or an underestimate of the actual distance? Give a reason for your answer.

It is an underestimate since all the chords lie under the curve.

**Key terms** Make sure you can write a definition for these key terms

chord    gradient    tangent

## 14 Non-linear real-life graphs

Learn the answers to the questions below, then cover the answers column with a
piece of paper and write as many as you can. Check and repeat.

| Questions | Answers |
|---|---|
| **1** For a distance–time graph, what does the gradient represent? | Speed. |
| **2** When you find the area under a curve, why is it an estimate? | You use trapeziums rather than following the exact curve. |
| **3** When is your answer an underestimate? | When the tops of the trapeziums are below the curve. |
| **4** On a velocity–time graph, what does the area under the curve represent? | Distance. |
| **5** How do you calculate the gradient between 2 points? | Draw a chord and calculate the gradient. |
| **6** How do you estimate the gradient at a point? | Draw a tangent and calculate the gradient. |
| **7** How do you calculate gradient? | Change in y over change in $x$. |
| **8** On a velocity time curve, what does the gradient of a tangent tell us? | The velocity at that point. |
| **9** What is the formula for the area of a trapezium? | $\frac{1}{2}(a+b)h$ |

Put paper here

## Previous questions

Now go back and use these questions to check your knowledge of previous topics.

| Questions | Answers |
|---|---|
| **1** How do you simplify a surd? | Look for a factor that is a square number. |
| **2** What is a linear graph? | A linear graph is a straight line which can be horizontal, vertical or sloping. |
| **3** What are the two algebraic methods you can use to solve simultaneous equations? | Elimination and substitution. |

Put paper here

## Exam-style questions

**14.1** The velocity–time graph shows the velocity, $v$ m/s, of a particle $t$ seconds after it is released.

   **(a)** Using four strips of equal width, estimate the area under the graph. **[3 marks]**

   **(b)** Explain what the area under the graph represents. **[1 mark]**

   **(c)** State whether your answer to part **(a)** is an overestimate or an underestimate and give a reason. **[1 mark]**

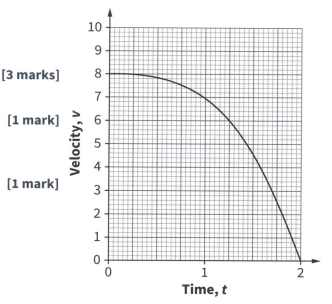

**14.2** The velocity–time graph shows the speed, $v$ m/s, of a ball as it rolls along a road for $t$ seconds.

   **(a)** Using three strips of equal length, estimate the total distance the ball travels in the first three seconds. **[3 marks]**

   **(b)** **(i)** Using three strips of equal length, estimate the total distance the ball travels in the second three seconds. **[3 marks]**

       **(ii)** State whether your answer to part **(b)(i)** is an overestimate or an underestimate and give a reason. **[1 mark]**

# Exam-style questions

**14.3** The graph shows the velocity, $v$, of a skateboarder over a period of time, $t$.

(a) Estimate the acceleration when $t = 5$      **[3 marks]**

(b) (i) Using two strips of equal width, estimate the area under the curve between $t = 15$ and $t = 20$      **[3 marks]**

        Give your answer to 2 s.f.
        You must show your working.

    (ii) State what your answer to part **(i)** represents.      **[1 mark]**

**14.4** David mows his lawn every few weeks. The graph shows the height of his lawn over a period of 12 weeks one summer.

(a) In which weeks does David cut the lawn?      **[3 marks]**

(b) Between which weeks does the grass grow at the slowest rate? **[1 mark]**

(c) What was the rate of growth of the grass, in cm/week, between weeks 5 and 7?      **[1 mark]**

**14.5** Here are the cross-sections of three swimming pools.

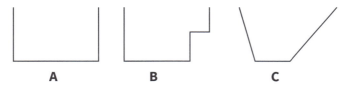

Each pool is filled with water from a hose at the same rate.

Match each pool to the graph of depth of water in the pool against time.

> **EXAM TIP**
> The steeper the graph, the faster the rate of change.

**[2 marks]**

**14.6** A ball is released from the top of a tower.
The distance–time graph shows information about its
fall to the ground.

**(a)** Calculate the average speed of the ball in the
first two seconds. **[2 marks]**

**(b)** Use the graph to estimate the speed of the ball
at time 1 second. **[3 marks]**

**EXAM TIP**

To find the average speed,
draw a chord. To find the
speed at a certain point,
draw a tangent.

 **14.7** Janusz goes on a train journey. The velocity–time graph shows the speed of
the train throughout the journey.

**(a)** Calculate the train's average acceleration during the first
30 minutes of the journey. **[2 marks]**

**(b)** Explain what is happening to the train's speed after 1.5 hours
of travel. **[1 mark]**

**(c)** Estimate the train's deceleration at 2 hours. **[3 marks]**

## Questions referring to previous content

 **14.8** Fill in the blanks in these sentences:

All exponential graphs of the form where $y = k^x$ is a positive constant, pass

through the point with coordinates _____ . When $k > 1$ the graph

will demonstrate exponential _____, and when $k < 1$ it demonstrates

exponential _____ . **[3 marks]**

**14.9** The function $y = f(x)$ undergoes a transformation given by $y = f(x - 2)$
Identify where the point $(7, 14)$ on the curve $y = f(x)$ will be when
translated by $y = f(x - 2)$ **[1 mark]**

# ⚙ Knowledge

## 15 Algebraic fractions, rearranging formulae with algebraic fractions, proof, functions and composite functions

## Algebraic fractions

An **algebraic fraction** is a fraction that contains at least one variable.

These are all examples of algebraic fractions: $\dfrac{3ab}{c^2}$ $\dfrac{x}{2}$ $\dfrac{3x^2 + 2x + 1}{5x - 4}$

Algebraic fractions can be simplified by cancelling common factors in the numerator and denominator.A

### Worked example

**1.** Simplify $\dfrac{24a}{36a^3}$

$= \dfrac{24a}{36a^3}$ ⟵ Simplify the multipliers by dividing by the HCF.

$= \dfrac{2a}{3a^3}$ ⟵ Simplify the '$a$'s by cancelling the indices.

$= \dfrac{2}{3a^2}$

**LINK**
To remind yourself about the rules of dividing indices, refer to Chapter 5.

**2.** Simplify $\dfrac{(b-1)(b+3)}{2(b-1)^2}$

$\dfrac{(b-1)(b+3)}{2(b-1)^2}$ ⟵ Expand the squared term.

$= \dfrac{(b-1)(b+3)}{2(b-1)(b-1)}$ ⟵ Cancel the factors that appear in both the numerator and denominator.

$= \dfrac{(b+3)}{2(b-1)}$ ⟵ Write the simplified fraction.

**3. (a)** Simplify fully $\dfrac{x^2 + 3x - 10}{x^2 - 25}$

$\dfrac{x^2 + 3x - 10}{x^2 - 25}$ ⟵ Factorise the numerator and denominator.

$= \dfrac{(x+5)(x-2)}{(x+5)(x-5)}$

$= \dfrac{(x-2)}{(x-5)}$ ⟵ Cancel the common factor (terms that appear in both the numerator and denominator).

**WATCH OUT** !
The remaining fraction will not simplify any further since there are no common factors of the numerator and denominator. Be careful, $x$ is not a common factor here since it isn't a multiplier, so you can't cancel it.

**(b)** Hence, solve $\dfrac{x^2 + 3x - 10}{x^2 - 25} = 2$

$\dfrac{(x-2)}{(x-5)} = 2$ ⟵ Replace the LHS with the simplified equation from part **(a)**.

$x - 2 = 2(x - 5)$

$x - 2 = 2x - 10$ ⟵ Rearrange the equation. Expand the brackets. Collect like terms to solve.

$x = 8$

## Multiplying algebraic fractions

To multiply algebraic fractions

- multiply the numerators together
- multiply the denominators together
- then cancel the common factors.

To divide by an algebraic fraction, multiply by its **reciprocal**.

**LINK**

To remind yourself about calculating with fractions, look back at Chapter 4.

---

**Worked example**

Simplify fully:

(a) $\dfrac{x^2 + 6x}{x^2 + 3x + 2} \times \dfrac{x^2 - 4x - 5}{3x^2}$

$= \dfrac{x(x+6)}{(x+1)(x+2)} \times \dfrac{(x+1)(x-5)}{3x^2}$

> Factorise the numerator and denominator of both fractions where possible.

$= \dfrac{x(x+6)(x+1)(x-5)}{3x^2 (x+1)(x+2)}$

> Multiply the two numerators and the two denominators.

$= \dfrac{(x+6)(x-5)}{3x(x+2)}$

> Simplify by cancelling common factors.

**WATCH OUT**

Do not expand brackets when multiplying algebraic fractions.

(b) $\dfrac{5x + 10}{x^2 - 9} \div \dfrac{5x^2}{x + 3}$

$= \dfrac{5x + 10}{x^2 - 9} \times \dfrac{x + 3}{5x^2}$

> Rewrite the calculation as a multiplication, by the reciprocal of the second fraction.

$= \dfrac{5(x+2)}{(x-3)(x+3)} \times \dfrac{x+3}{5x^2}$

> Factorise where possible.

**WATCH OUT**

You can cancel before or after multiplying, but be careful to only cancel common factors of a numerator and a denominator.

$= \dfrac{5(x+2)\cancel{(x+3)}}{5x^2 (x-3)\cancel{(x+3)}}$

> Multiply the fractions together.

$= \dfrac{x+2}{x^2 (x-3)}$

> Cancel common factors.

---

## 15 Algebraic fractions, rearranging formulae with algebraic fractions, proof, functions and composite functions

## Adding or subtracting algebraic fractions

Algebraic fractions can be added or subtracted.

> **Worked example**
>
> Simplify fully $\dfrac{x+1}{x+3} + \dfrac{x-1}{x^2+4x+3}$
>
> $= \dfrac{x+1}{x+3} + \dfrac{x-1}{(x+1)(x+3)}$ — Factorise where possible.
>
> $= \dfrac{(x+1)(x+1)}{(x+1)(x+3)} + \dfrac{x-1}{(x+1)(x+3)}$ — Find the **LCM** of the denominators and rewrite as equivalent fractions. Remember to multiply the numerator by the same factor as the denominator.
>
> $= \dfrac{(x+1)(x+1)+(x-1)}{(x+1)(x+3)}$
>
> $= \dfrac{x^2+2x+1+x-1}{(x+1)(x+3)}$ — Add the fractions.
>
> $= \dfrac{x^2+3x}{(x+1)(x+3)}$ — Expand the brackets in the numerator.
>
> $= \dfrac{x(x+3)}{(x+1)(x+3)}$ — Collect like terms to simplify.
>
> $= \dfrac{x}{x+1}$ — Factorise the numerator.
>
> — Cancel common factors.

> **WATCH OUT** (!)
>
> Always factorise and find the LCM of the denominators. If you cross-multiply by the denominators without factorizing and using the LCM, the algebra can get quite tricky.

## Proof

An **equation** is only true for certain values.
For example, $x+5=8$ is only true for $x=3$.

An **identity** is true for all values.
For example, $2x+3x=5x$ is true for all values of $x$, so you can write $2x+3x \equiv 5x$.

If you are asked to **prove** a result, it will always be an identity.

To prove an identity, start with the expression on one side and use algebraic techniques, such as expanding and simplifying, until you reach the expression on the other side.

> **REVISION TIP**
>
> Equation = for certain values
>
> identity ≡ for all values

> **Worked example**
>
> 1. Prove that $(a+b)^2 - (a-b)^2 \equiv 4ab$
>
> $\text{LHS} \equiv (a+b)^2 - (a-b)^2$ — Start with the left-hand side (LHS). Expand the brackets.
>
> $\equiv (a^2+2ab+b^2) - (a^2-2ab+b^2)$
>
> $\equiv a^2+2ab+b^2-a^2+2ab-b^2$ — Collect like terms.
>
> $\equiv 4ab$ — Compare to the right-hand side (RHS).
>
> $\equiv \text{RHS}$

## Proof

> **Worked example**
>
> Prove that the difference between the squares of any two consecutive odd numbers is always a multiple of 8.
>
> Let $n$ be an integer. •————— If $n$ is an integer, you can use $2n$ to represent even numbers and $2n + 1$ to represent odd numbers.
>
> Then $2n + 1$ and $2n + 3$ are consecutive odd numbers. •——— Write an expression to represent the statement.
>
> difference between the squares $= (2n + 3)^2 - (2n + 1)^2$
>
> $= (4n^2 + 12n + 9) - (4n^2 + 4n + 1)$ •——— Expand the brackets.
>
> $= 8n + 8$ •——— Collect like terms to simplify.
>
> $= 8(n + 1)$ •——— To show an expression is a multiple of a number, write it with the number as a factor.

If a result is *not* true, you only need to find one example to disprove the result. This is called a counterexample.

> **Worked example**
>
> Anis says that the difference between two square numbers is always even.
> Give one example to show that Anis is not correct.
>
> $2^2 = 4$ and $3^2 = 9$
>
> $9 - 4 = 5$, which is an odd number. This disproves Anis's claim.

## Functions

**Functions** can be labelled as $f(x)$, $g(x)$, and so on. For example, $f(x) = x^2$.

You can **evaluate** a function by substituting a value in place of $x$.

> **Worked example**
>
> $f(x) = 4x^2 + 2x - 1$
>
> **(a)** Work out $f(-3)$
>
> $f(-3) = 4(-3)^2 + 2(-3) - 1$ •——— Substitute $-3$ for $x$.
>
> $= 36 - 6 - 1$ •——— Expand the brackets and simplify.
>
> $= 29$
>
> **(b)** Work out values of $x$ for which $f(x) = 1$
>
> $4x^2 + 2x - 1 = 1$ •——— Set the function equal to 1
>
> $4x^2 + 2x - 2 = 0$ •——— Subtract 1 from both sides to get a quadratic equal to 0 and solve.
>
> $2x^2 + x - 1 = 0$
>
> $(2x - 1)(x + 1) = 0$ •——— Factorise the equation.
>
> $x = \frac{1}{2}$ or $x = -1$ •——— Then set each factor equal to 0 to find the two values of $x$.

> **LINK**
> To remind yourself about the order of operations, see Chapter 1.

> **LINK**
> To remind yourself about solving quadratic equations, see Chapter 9.

# ⚙ Knowledge

## 15 Algebraic fractions, rearranging formulae with algebraic fractions, proof, functions and composite functions

### Composite functions

Functions can be combined to form **composite functions**, such as fg($x$).

The function closest to ($x$) is carried out first.

For example, if $f(x) = x^2$ and $g(x) = x + 2$

$fg(x) = f(x + 2)$ ⟶ fg($x$) means function g followed by function f.

$\quad = (x + 2)^2$

$gf(x) = g(x^2)$ ⟶ gf($x$) means function f followed by function g.

$\quad = x^2 + 2$

---

**Worked example**

$f(x) = 2x + 1, g(x) = 5 - x^2$

**(a)** Work out gf(2).

$\quad f(2) = 2(2) + 1 = 5$ ⟵ Substitute 2 into f($x$).

$\quad gf(2) = g(5)$

$\quad\quad = 5 - 5^2$ ⟵ Substitute the answer into g($x$).

$\quad\quad = -20$

**(b)** Write a simplified expression for fg($x$).

$\quad fg(x) = f(5 - x^2)$ ⟵ Substitute the expression for g.

$\quad\quad = 2(5 - x^2) + 1$ ⟵ Substitute the expression for $x$.

$\quad\quad = 11 - 2x^2$ ⟵ Expand and simplify.

**(c)** Solve the equation fg($x$) = g($x$).

$\quad 11 - 2x^2 = 5 - x^2$ ⟵ Write the substituted expressions as equal.

$\quad\quad x^2 = 6$ ⟵ Collect like terms.

$\quad x = \sqrt{6}$ or $x = -\sqrt{6}$ ⟵ Rearrange to solve.

---

### Inverse functions

The **inverse** of a function reverses its effect.

The inverse of f($x$) is written f$^{-1}$($x$).

The inverse of a function can be found using a function machine. For example, the function machine for $f(x) = 3x - 7$ is:

To find f$^{-1}$($x$), replace each operation with its inverse and work from right to left:

$$f^{-1}(x) = \frac{x + 7}{3}$$

## Inverse functions

So $f^{-1}(x) = \dfrac{x+7}{3}$

The inverse can also be found algebraically by rearranging the formula to make $x$ the **subject**.

---

**Worked example**

$f(x) = 9 - 5x$ for all values of $x$.

**(a)** Work out an expression for $f^{-1}(x)$.

$y = 9 - 5x$ ← Write the formula, replacing $f(x)$ with $y$.

$5x = 9 - y$

$x = \dfrac{9 - y}{5}$ ← Rearrange the formula to make $x$ the subject.

$f^{-1}(x) = \dfrac{9 - x}{5}$

Replace:
- $x$ with $f^{-1}(x)$
- $y$ with $x$.

**(b)** Solve the equation $f(x) = f^{-1}(x)$.

$9 - 5x = \dfrac{9 - x}{5}$ ← Write an equation using the expressions for $f(x)$ and $f^{-1}(x)$.

$45 - 25x = 9 - x$

$36 = 24x$

$x = \dfrac{36}{24}$ ← Rearrange and collect like terms to solve.

$= 1.5$

---

**Worked example**

$f(x) = 5x - 2$, $g(x) = x^3 + 1$

The function h is such that $h(x) \equiv gf(x)$.

**(a)** Work out an expression for $h^{-1}(x)$.

$h(x) = g(5x - 2)$ ← Write $h(x)$ using the functions given.

$= (5x - 2)^3 + 1$

$y = (5x - 2)^3 + 1$

$y - 1 = (5x - 2)^3$ ← Rearrange to find the inverse.

$\sqrt[3]{y - 1} = 5x - 2$

$2 + \sqrt[3]{y - 1} = 5x$

$x = \dfrac{1}{5}\left(2 + \sqrt[3]{y - 1}\right)$

$h^{-1}(x) = \dfrac{1}{5}\left(2 + \sqrt[3]{x - 1}\right)$ ← Replace $x$ with $h^{-1}(x)$, and replace $y$ with $x$.

**(b)** Given that $h(a) = 9$, find the value of $a$.

If $h(a) = 9$ ← Substitute the value into the equation and solve.

$a = h^{-1}(9)$

$= \dfrac{1}{5}\left(2 + \sqrt[3]{9 - 1}\right)$

$= 0.8$

---

**Key terms**

Make sure you can write a definition for these key terms.

algebraic fraction   composite function   equation   evaluate
inverse   inverse function   identity   lowest common multiple
proof   reciprocal   subject

## 15 Algebraic fractions, rearranging formulae with algebraic fractions, proof, functions and composite functions

Learn the answers to the questions below, then cover the answers column with a piece of paper and write as many as you can. Check and repeat.

## Questions | Answers

| | Questions | | Answers |
|---|---|---|---|
| 1 | How do you simplify algebraic fractions? | Put paper here | Look for common factors to cancel – you may need to factorise the numerator and/or the denominator first. |
| 2 | How do you add or subtract algebraic fractions? | | Find a common denominator. You may need to factorise first. |
| 3 | What does 'make $x$ the subject of the equation' mean? | Put paper here | Rearrange the equation into the form $x = \ldots$ |
| 4 | What symbol does an identity include? | | The identity symbol, $\equiv$. |
| 5 | If $n$ is an integer, how could you represent an even number? | Put paper here | $2n$ |
| 6 | If $n$ is an integer, how could you represent an odd number? | | $2n + 1$ or $2n - 1$ (other answers are possible). |
| 7 | How do you represent a function? | Put paper here | $f(x)$ or $g(x)$ (other letters can be used). |
| 8 | What is a composite function? | | Two or more functions combined. |
| 9 | In $fg(x)$, which function happens first? | | $g(x)$ |
| 10 | What are the three steps you need to follow to find the inverse of a function $f(x)$? | Put paper here | 1. Write in the form $y = \ldots$<br>2. Rearrange to make $x$ the subject.<br>3. Replace $y$ with $x$ and $x$ with $y$, and write as $f^{-1}(x) = \ldots$ |

## Previous questions

Now go back and use these questions to check your knowledge of previous topics.

## Questions | Answers

| | Questions | | Answers |
|---|---|---|---|
| 1 | What is an integer? | | A whole number. |
| 2 | What is a surd? | | An irrational square root. |
| 3 | What is an expression? | Put paper here | A collection of letters and numbers which cannot be solved (no equals or inequality sign). |
| 4 | How do you solve two inequalities that require you to show the solutions on a number line? | | Solve each inequality separately then combine the solutions. |
| 5 | What are the two algebraic methods you can use to solve simultaneous equations? | | Elimination and substitution. |

## Exam-style questions

 **15.1** **(a)** Expand and simplify fully $(x-1)(x-2)(x-3)$. **[3 marks]**

**(b)** Hence, or otherwise, prove that $n^4 - 6n^3 + 11n^2 - 6n$ is always even for all integer values of $n > 3$. **[2 marks]**

 **15.2** Simplify fully

**(a)** $\dfrac{4x^2 - 12x}{2x}$ **[2 marks]**

**(b)** $\dfrac{x^2 - x - 2}{x^2 - 6x + 8}$ **[3 marks]**

 **15.3** Simplify fully **[3 marks]**

**(a)** $\dfrac{3x - 4}{6x^2 + 7x - 20}$

**(b)** Hence, solve the equation $\dfrac{3x - 4}{6x^2 + 7x - 20} = 1$ **[2 marks]**

 **15.4** Solve $\dfrac{1}{x - 3} = \dfrac{x}{x + 5}$ **[3 marks]**

 **15.5** Simplify fully $\dfrac{2x - 8}{3x - 15} \times \dfrac{x - 5}{x - 3}$ **[3 marks]**

**15.6** Simplify fully $\dfrac{x}{x^2 + 2x - 35} \div \dfrac{6x^3}{3x + 21}$ **[4 marks]**

 **15.7** $x^2 : (x - 1) = 4 : 1$

Work out the value of $x$. **[4 marks]**

**15.8** Write as a single fraction in its simplest form.

$\dfrac{4}{x + 1} + \dfrac{3}{x + 2}$ **[3 marks]**

**15.9** Write as a single fraction in its simplest form $\dfrac{2}{(x+4)} - \dfrac{3}{(x-2)}$ **[3 marks]**

 **15.10** $f(x) = x^2 - 2x$

    **(a)** Work out the value of

        **(i)** $f(1)$ **[2 marks]**

        **(ii)** $f(-5)$. **[2 marks]**

    **(b)** Work out the values of $x$ for which

        **(i)** $f(x) = 15$ **[3 marks]**

        **(ii)** $f(x) = 4x$. **[3 marks]**

 **15.11** $f(x) = 3x - 1$ and $g(x) = \dfrac{1}{x}$

    Work out the value of

    **(a)** $gf(1)$ **[3 marks]**

    **(b)** $fg\left(\dfrac{1}{2}\right)$ **[3 marks]**

 **15.12** $g$ is a function such that

    $g(x) = x^2 - 4, x \geq 0$ **[4 marks]**

    Work out the inverse function $g^{-1}(x)$.

 **15.13** Prove that $(m+n)^2 + (m-n)^2 \equiv 2(m^2 + n^2)$. **[3 marks]**

 **15.14** Prove algebraically that the sum of five consecutive integers is divisible by 5. **[3 marks]**

 **15.15** Show that $\dfrac{2x^2 + 3x}{9x^2 - 4} \times \dfrac{9x^2 + 42x + 24}{x^2 + 4x}$ simplifies to $\dfrac{a(bx + c)}{dx + e}$,

    where $a, b, c, d$ and $e$ are integers. **[4 marks]**

**15.16** Express $\dfrac{2}{x-1} + \dfrac{3}{x-2} + \dfrac{3}{2x}$ as a single fraction in its simplest form. **[4 marks]**

**15.17**  $f(x) = \frac{1}{x}$      $g(x) = \frac{2}{x^2}$

     **(a)**    Work out fg(1).                             **[2 marks]**

     **(a)**    Work out gf(x).                            **[2 marks]**

**15.18**  $f(x) = 2x + 1$    $g(x) = x^2$

     **(a)**    Work out ff(x).                             **[2 marks]**

     **(b)**   $f^{-1}(x) = fg(x)$
           Show that $4x^2 - x + 3 = 0$.                   **[6 marks]**

**15.19**  Prove that the sum of two consecutive odd numbers is always a multiple of 4.                             **[3 marks]**

**Questions referring to previous content**

**15.20** Multiply these numbers and write your answer in standard form:

       $27.2 \times 10^6 \times 1.75 \times 10^{-3}$                   **[1 mark]**

**15.21**  Factorise $-12xy^2 + 15xyz - 3x^2y + 21ax^3y^2$             **[2 marks]**

## 16 Ratio

## Ratios: an overview

A **ratio** describes how things are split into **parts**.

The ratio of black beads to red beads on this bracelet is $3:2$.

Ratio problems can sometimes be solved by 'scaling up' a ratio. You do this by multiplying both sides by the same number.

Ratios compare parts to parts; **fractions** compare parts to the whole. For example, if the ratio of adults to children is $1:2$ then $\frac{1}{3}$ of the total are adults and $\frac{2}{3}$ are children.

Ratios can also have three or more parts, for example $a:b:c$.

A ratio can be simplified fully by dividing both sides by their highest common factor (HCF).

**Worked example**

Simplify the ratio $49:28$

$\left(\frac{49}{7}\right):\left(\frac{28}{7}\right) = 7:4$

> The HCF of 49 and 28 is 7, so divide both by 7.

**LINK**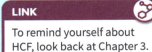

To remind yourself about HCF, look back at Chapter 3.

## Using ratio

**WATCH OUT** (!)

These next two questions look very similar. Make sure you focus on the **information** and **instructions** given.

**Worked example**

Ben, Sima, and Lisa share some money in the ratio $3:4:5$. Lisa gets £30. How much does Sarah get?

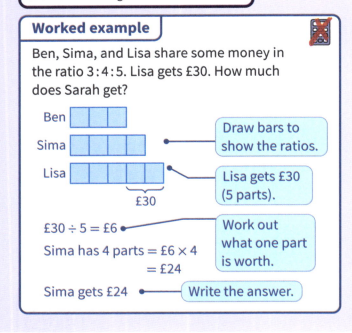

Ben

Sima

Lisa

£30

> Draw bars to show the ratios.

> Lisa gets £30 (5 parts).

$£30 \div 5 = £6$

> Work out what one part is worth.

Sima has 4 parts $= £6 \times 4$
$= £24$

Sima gets £24

> Write the answer.

**Worked example**

Ben, Sima, and Lisa share some money in the ratio $3:4:5$. Lisa gets £30 more than Ben. How much does Sima get?

Ben

Sima

Lisa

£30

> Draw bars to show the ratios.

> Lisa gets £30 (2 parts) more than Ben.

$£30 \div 2 = £15$

> Work out what one part is worth.

Sima has 4 parts $= £15 \times 4$
$= £60$

Sima gets £60

> Write the answer.

# Using ratio

## Worked example

In a class of children, the ratio of 8-year-olds to 9-year-olds is 2:3.

25% of the 8-year-olds are left-handed.

$\frac{1}{3}$ of the 9-year-olds are left-handed.

What percentage of the class is right-handed?

8-year-olds

9-year-olds

Draw bars to show the ratios.

8-year-olds  L

9-year-olds  L

Split the 8-year-olds boxes into 4 so you can show 25% are left-handed. Mark $\frac{1}{3}$ of the 9-year-olds boxes to show left-handed.

8-year-olds  L R R R

9-year-olds  L L R R R R

Make all the boxes the same size. Mark and count the right-handed boxes.

$\frac{7}{10}$ are right-handed = 70%

Write the answer.

# Ratios and scale

A **scale** can be written as a ratio, such as 1:400, or as a sentence.

You multiply the length on the drawing to get the real-life length.

You divide the real-life length to get the length on the drawing.

## Worked example

A drawing of a tower is 1.4 cm tall. It is drawn to a scale of 1:400

How tall is the tower in real life?

| drawing | real life |
|---------|-----------|
| 1 | 400 |
| 1 cm | 400 cm |
| 1 cm | 4 m |

Make a table.

Write the scale.

Put in the units.

Size of drawing = 1.4 cm

Change to better units.

Actual size = 1.4 × 4

= 5.6 m

## Worked example

On a map, 2 cm represents 3 km.

Write the scale of the map as a ratio in the form 1:$n$

| map | real |
|-----|------|
| 2 cm | 3 km |
| 2 cm | 3000 m |
| 2 cm | 300 000 cm |
| 1 | 150 000 |

Simplify.

# ⚙ Knowledge

## 16 Ratio

## Problem solving with ratios

### Worked example

Oskar mixes red and yellow paint in the ratio 2 : 3

1. Oskar needs 240 ml of red paint to paint a room.

   How much yellow paint does he need?

   [Draw a **bar model** in the ratio 2 : 3]

   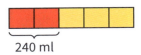

   240 ml

   Oskar needs 360 ml of yellow paint.

   [240 ml is 2 parts so 1 part
   $= 240 \div 2 = 120$
   Yellow paint is 3 parts, so 3 parts
   $= 120 \times 3 = 360$]

2. Oskar needs 15 litres of the mixed paint to paint his house.

   How much red paint does he need?

   [1 part $= 15 \div 5 = 3$
   Red $= 2$ parts $= 3 \times 2 = 6$]

   15 litres

   Oskar needs 6 litres of red paint.

## Using fractions and percentages with ratio

If $a : b = 3 : 5$, then $\frac{a}{b} = \frac{3}{5}$

If $b$ is 20% less than $a$, then $a : b = 100 : 80$, which is 5 : 4.

If $2a = 3b$, then $a : b = \frac{3}{2} : 1$, which is 3 : 2 (be careful here: $a$ is greater than $b$).

### Worked example

A bag contains green and blue beads in the ratio 3 : 4.

Write an equation for $y$ in terms of $x$ to show the relationship between the number of green beads ($y$) and the number of blue beads ($x$).

$\frac{3}{4} : 1$   [Write the ratio in the form $n : 1$
Turn this into an equation.]

$y = \frac{3}{4}x$   [You can substitute in some values to check.]

## Using fractions and percentages with ratio

In questions where the ratio changes:

- Write an equation for each situation using fractions.
- Form linear equations.
- Solve the equations simultaneously.

> **LINK**
>
> To remind yourself about solving simultaneous equations, look back at Chapter 11.

**Worked example**

A bag contains red and blue counters in the ratio $1:5$.

Three red counters are added to the bag. The ratio of red to blue counters is now $1:4$.

How many blue counters are there in the bag?

Let $x$ be the number of red counters at the start.

> Write an equation for each situation.

At the start, there are:

- $x$ red counters
- $5x$ blue counters
- $6x$ counters altogether

Let $y$ be the number of red counters at the end.

At the end, there are:

- $y$ red counters
- $4y$ blue counters
- $5y$ counters altogether

3 red counters are added to the bag, so

$y = x + 3$     (equation 1)

Total increase in counters is 3, so

$5y = 6x + 3$     (equation 2)

> Write the equations as simultaneous equations.

$$5(x + 3) = 6x + 3$$
$$5x + 15 = 6x + 3$$

> Expand the brackets and solve to find $x$.

$$15 - 3 = 6x - 5x$$
$$12 = x$$

$$y = x + 3$$
$$y = 12 + 3$$

> Use substitution to find $y$.

$$y = 15$$

blue counters at end $= 4y$
$$= 4 \times 15$$

> Multiply to find number of blue counters at the end.

$$= 60 \text{ blue counters.}$$

Let $x$ be the number of red counters at the start.
At the start, there are: ~ $x$ red counters ~ $5x$ blue counters
~ $6x$ counters altogether.

**Key terms**    Make sure you can write a definition for these key terms    | bar model   fraction   parts   ratio   scale |

## 16 Ratio

Learn the answers to the questions below, then cover the answers column with a piece of paper and write as many as you can. Check and repeat.

## Questions | Answers

| | Questions | Answers |
|---|---|---|
| 1 | What is a ratio used for? | To compare two or more quantities in relation to each other. |
| 2 | How do we write a ratio in the form $1 : n$? | Divide both values to make the first part 1. |
| 3 | How do we know when a ratio is written in its simplest form? | The only remaining common factor of each of the parts of the ratio is 1. |
| 4 | To simplify a ratio fully, what do you need to divide all the parts in the ratio by? | The highest common factor of the parts. |
| 5 | The ratio of adults to children in a group is $4 : 7$. Explain how to express the number of adults as a fraction of the whole group. | There are 4 parts representing adults and $4 + 7 = 11$ parts in total, so the fraction of adults is $\frac{4}{11}$. |
| 6 | How do you scale up a ratio? | Multiply each part by the same number. |
| 7 | True or false?<br>A ratio can only have two parts. | False, ratios can have more than two parts, e.g. $3 : 4 : 6$. |
| 8 | On a scale drawing the ratio is $1 : n$. How do you find the length on a drawing? | Divide the real-life length by $n$. |
| 9 | What does a ratio of $1 : 100$ mean on a scale drawing? | 1 unit in the drawing is equal to 100 units in real life. |
| 10 | In a particular city the ratio of red cars to blue cars is r : b. What is the ratio of blue cars to red cars? | b : r |

*Put paper here*

## Previous questions

Now go back and use these questions to check your knowledge of previous topics.

## Questions | Answers

| | Questions | Answers |
|---|---|---|
| 1 | How do you divide by a fraction? | Multiply by the reciprocal of the fraction. |
| 2 | What is the highest common factor of two or more numbers? | The largest number that is a factor of all the numbers. |
| 3 | How do you convert a fraction into a decimal? | Divide the numerator by the denominator. |
| 4 | What is an improper fraction? | A fraction where the numerator is bigger than the denominator. |
| 5 | What is the midpoint formula for finding the midpoint of two points $(x_1, y_1)$ and $(x_2, y_2)$? | $\text{midpoint} = \left( \frac{x_1 + x_2}{2}, \frac{y_1 + y_2}{2} \right)$ |

*Put paper here*

## Exam-style questions

**16.1** The lengths of the sides of a right-angled triangle are in the ratio 5 : 12 : 13

The length of the shortest side is 2 cm.

Work out the perimeter of the triangle. **[3 marks]**

**16.2** The ratio of pencils to erasers is 7 : 2.

There are 90 more pencils than erasers.

Work out the number of pencils. **[2 marks]**

 **16.3** The points $A$, $B$, $C$ and $D$ form a straight line $ABCD$.

The ratio of the length of $AB$ to length $BC$ is 6 : 5

The ratio of length $BC$ to length $CD$ is 10 : 13

The total length of $AD$ is 105 cm.

Work out the length of $BC$. **[3 marks]**

**16.4** Kayleigh is two years younger than Hayley.

Bailey is twice as old as Kayleigh.

The sum of their three ages is 38.

Find the ratio of Bailey's age to Hayley's age to Kayleigh's age. **[3 marks]**

 **16.5** Carl and Frida have 44 cards.

Carl has 20% more cards than Frida.

Work out the number of cards they each have. **[3 marks]**

 **16.6** In a traffic survey the ratio of blue cars to red cars is 2 : 1

In a follow-up survey, the number of blue cars has increased by 50% and the number of red cars has increased by 25%.

What is the new ratio of blue cars to red cars?

Write your answer in the form $a : b$, where $a$ and $b$ are integers. **[3 marks]**

**16.7** The ratio of integer $x$ to integer $y$ is 3 : 2.
When $x$ is increased by 8 and $y$ is halved, the ratio becomes 4 : 1

Work out the values of $x$ and $y$. **[5 marks]**

**16.8** James has some copper coins and some silver coins.

Each coin is either large or small.

The ratio of copper coins to silver coins is 4 : 3

The ratio of small copper coins to large copper coins is 3 : 7

The ratio of small silver coins to large silver coins is 1 : 2

Work out what fraction of all of the coins are small coins. **[5 marks]**

**16.9** Given that $(2x - 5) : 6 = 1 : (6 - x)$,

work out the possible values of $x$. **[4 marks]**

> **EXAM TIP**
>
> $a : b = c : d$ then $\dfrac{a}{b} = \dfrac{c}{d}$

**16.10** A world cruise costs £15 285 plus VAT at 20%.

Judith pays a deposit followed by 10 equal payments of £1384.20

Work out the ratio

deposit : total monthly payments

Give your answer in its simplest form. **[4 marks]**

**16.11** Light travels $4.8555 \times 10^9$ km in 4.5 hours.

Sound travels $3.7044 \times 10^4$ km in 3 hours.

Work out the ratio

speed of light : speed of sound

Give your answer in the form $n : 1$, where $n$ is correct to 3 significant figures. **[4 marks]**

**16.12** Pranay, Quinten, Raha and Stephanie share £425.

The ratio of the amount Pranay gets to the amount Raha gets is 4 : 5

Quinten gets 1.25 times the amount Pranay gets.

Stephanie gets 2.2 times the amount Quinten gets.

Work out the amount of money that Stephanie gets. **[4 marks]**

**16.13** A delivery consists of 560 pieces of fruit. Each piece of fruit is either a banana, an apple, an orange or a pear. $\frac{3}{8}$ of the fruit are oranges. 15% of the fruit are bananas. The ratio of the number of apples to the number of pears is 8 : 11.

Work out how many pears there are in the delivery. **[5 marks]**

**16.14** Given that $3x^2 : (5x + 4) = 2 : 1$
work out the possible values of $x$. **[4 marks]**

**16.15** Deshawn, Amara and Harper share some marshmallows in the ratio $3 : 2 : 5$
Harper gives five of her marshmallows to Amara, one of which she eats. The
ratio of marshmallows is then $4 : 4 : 5$

Work out how many marshmallows each person had to begin with. **[3 marks]**

**Questions referring to previous content**

 **16.16** Work out $3\frac{2}{3} \div 2\frac{1}{2}$

Give your answer as a mixed number. **[3 marks]**

 **16.17** Simplify fully $\frac{x^2 + 7x + 10}{x^2 + 2x - 15} \times \frac{x^2 + x - 12}{x^2 + 2x}$. **[4 marks]**

## 17 Compound measures and multiplicative reasoning

### Measures

Convert between metric units by multiplying or dividing by powers of 10.

**milli-** = one thousandth

**centi-** = one hundredth

**kilo-** = one thousand

| length 1 m | area 1 m² | volume 1 m³ |
|---|---|---|
| 100 cm | 10 000 cm² | 1 000 000 cm³ |

#### Worked example

A sphere has a volume of 20 cm³ and a surface area of 36 cm².

1.  Write the surface area in mm²

    Surface area = 36 × 10² = 3600 mm²  •————  To convert from cm² to mm², multiply by 10²

2.  Write the volume in mm³.

    Volume = 20 × 10³ = 20 000 mm³  •————  To convert from cm³ to mm³, multiply by 10³

### Compound measures

A **compound measure** links two measurements. For example, a rate of pay could be £11 per hour and a rate of flow of water could be 3 litres per minute.

You need to know the formulae for these compound measures:

$$\text{speed} = \frac{\text{distance}}{\text{time}} \qquad \text{density} = \frac{\text{mass}}{\text{volume}} \qquad \text{pressure} = \frac{\text{force}}{\text{area}}$$

#### Worked example

Imani is buying a car. She compares the maximum speeds of two cars.

car 1: top speed = 195 km/h

car 2: top speed = 52 m/s

Which car has the highest top speed?

To compare top speeds, we need to convert the speeds to the same units. Here we will convert car 1 to m/s, but you could convert car 2 to km/h.

195 × 1000 = 195 000 m/h  •————  × 1000 to convert to m/h.

$\dfrac{195\,000}{60 \times 60} = 54.17$ m/s  •————  Divide by 60 to convert from m/h to m/min, and divide by 60 again to convert to m/s.

So car 1 has the higher top speed.

## Compound measures

### Worked example

A solid piece of wood has mass 1.8 kilograms density 2.4 g/cm³
Work out the volume of the wood.

$$\text{density} = \frac{\text{mass}}{\text{volume}}$$ — Write the formula linking density, mass and volume.

$$\text{volume} = \frac{\text{mass}}{\text{density}}$$ — Rearrange to make volume the subject.

$$1.8 \text{ kg} = 1800 \text{ g}$$ — Convert kg to g.

$$\text{volume} = \frac{1800}{24}$$
$$= 750 \text{ cm}^3$$ — Substitute to find volume.

## Direct proportion

When two values are in **direct proportion**, they will increase and decrease at the same rate.

For example, the number of apples and the cost of the apples are in direct proportion. If one apple costs 30p, then:

| apples | cost |
|--------|------|
| 1 | 30p |
| 2 | 60p |
| 7 | 140p |

To solve direct proportion problems, you can use a common factor, or the **unitary method**:

- Divide to calculate the value of 1 unit.
- Multiply, if necessary, to find the value required.

The symbol for 'proportional to' is $\propto$.

### Worked example: Common factors

**1.** A pack of 20 biscuits costs £1.50 and a pack of 35 of the same biscuits costs £2.80. Which is the better value?

| small pack | |
|------|--------|
| 20 | £1.50 |
| 5 | £0.375 |

÷4 ... ÷4

| large pack | |
|------|--------|
| 35 | £2.80 |
| 5 | £0.40 |

÷7 ... ÷7

5 is a common factor.

The smaller pack is better value.

**2.** A recipe for 12 portions of cheesecake requires 30 biscuits. How many biscuits are required for 32 portions of cheesecake?

| portions | biscuits |
|----------|----------|
| 12 | 30 |
| 4 | 10 |
| 32 | 80 |

÷3 ... ÷3
×8 ... ×8

80 biscuits are required for 32 portions.

### Worked example: Unitary method

A pack of 240 teabags costs £2.25 and a pack of 150 of the same teabags costs £1.10.

Which is the better value?

| 240 pack | | 150 pack | |
|----------|-------|----------|-------|
| bags | price | bags | price |
| 240 | £2.25 | 150 | £1.10 |
| 1 | £0.009375 | 1 | £0.0073 |

Divide the price by the number of bags to get the price per unit.

The 150 pack is better value. — Compare the price per unit; lowest is best value.

# Knowledge

## Graphs of direct proportion

If $x$ is directly proportional to $y$, then $y = kx$, where $k$ is the constant or proportionality.

A graph showing two variables in direct proportion is a straight line through the origin with gradient $k$.

**LINK**

To remind yourself about straight-line graphs, look back at Chapter 6.

**Worked example**

The graph shows the exchange rate between pounds (£$y$) and dollars ($\$x$) on a given day.

(a) Work out an equation for $y$ in terms of $x$.

gradient $= \dfrac{8}{10} = 0.8$, so $y = 0.8x$

(b) Use your equation to convert $15 to pounds.

$y = 0.8 \times 15 = 12$ so £12.

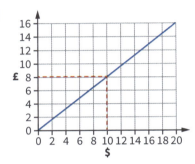

## Equations for direct proportion

$y$ can also be directly proportional to a function of $x$, for example, $y = kx^2$ or $y = k\sqrt{x}$

**Worked example**

$b$ is directly proportional to $a^2$

$b = 20$ when $a = n$

Show that when $a = 3n$, $b = 180$

$b \propto a^2$ so $b = ka^2$  —— Write an equation with $k$ as the constant of proportionality.

$20 = kn^2$ —— Substitute $b = 20$, $a = n$

$k = \dfrac{20}{n^2}$ —— Rearrange to give an expression for $k$.

$b = \left(\dfrac{20}{n^2}\right) a^2$

$\phantom{b} = \dfrac{20a^2}{n^2}$ —— Substitute $k = \dfrac{20}{n^2}$ into $b = ka^2$

$\phantom{b} = \dfrac{20(3n)^2}{n^2}$ —— Substitute $a = 3n$ into $b = \dfrac{20a^2}{n^2}$

$\phantom{b} = \dfrac{180(n)^2}{n^2}$

$\phantom{b} = 180$

# Inverse proportion

Two values are in **inverse proportion** if, when one *increases*, the other *decreases* at the same rate.

For example, the number of taps and time taken to fill a sink are in inverse proportion.

| time | taps |
|------|------|
| 6 mins | 1 |
| 3 mins | 2 |
| 1.5 mins | 4 |

÷2 (6 mins → 3 mins → 1.5 mins) ×2 (1 → 2 → 4)

Notice that the product of the two values is constant.

**Worked example**

It takes 8 cleaners 3 hours to clean a block of offices.

1. How long does it take 6 cleaners to clean the offices?

| cleaners | hours | cleaner-hours |
|----------|-------|---------------|
| 8 | 3 | $8 \times 3 = 24$ |
| 6 | $\frac{24}{6} = 4$ | 24 |

> Work out how many cleaner-hours the offices require.

> Work out how many hours are needed if there are 6 cleaners.

It takes 6 cleaners 4 hours to clean the offices.

2. How many cleaners are needed to clean the offices in 1.5 hours?

| cleaners | hours | cleaner-hours |
|----------|-------|---------------|
| 8 | 3 | 24 |
| 16 | 1.5 | 24 |

×2 (8 → 16), ÷2 (3 → 1.5)

> Work out how many cleaners are needed if there are 1.5 hours. $(16 \times 1.5 = 24)$

16 cleaners are needed to clean the offices in 1.5 hours.

# Graphs of inverse proportion

A graph showing inverse proportion will lie close to the $x$ and $y$ axes, and curve away from the origin.

If $x$ is inversely proportional to $y$, then $y = \dfrac{k}{x}$ where $k$ is any number and is called the **constant of proportionality**.

**LINK**

To remind yourself about reciprocal graphs, look back at Chapter 13.

# Equations for inverse proportion

It is also possible for $y$ to be inversely proportional to a function of $x$.
For example,

$y = \dfrac{k}{\sqrt{x}}$, here $y$ is inversely proportional to the square root of $x$.

**Worked example**

$y$ is inversely proportional to $x^2$

$y = 2$ when $x = 3$

Work out the value of $y$ when $x = 4$

$y = \dfrac{k}{x^2}$

> Write an equation connecting $x$ and $y$.

$2 = \dfrac{k}{9}$

$k = 18$

When $x = 4$,

> Substitute $x$ and $y$ and rearrange to find $k$.

$y = \dfrac{18}{4^2}$

$= \dfrac{9}{8}$

> Substitute $k$ and $x$ into the equation and solve for $y$. Remember to simplify the fraction.

## Problem solving with proportion

**Worked example**

$m$ is directly proportional to $v^3$

$m = 24$ when $v = 2$

$v$ is inversely proportional to $n^2$

$v = 4$ when $n = 10$

Work out an equation connecting $m$ and $n$.

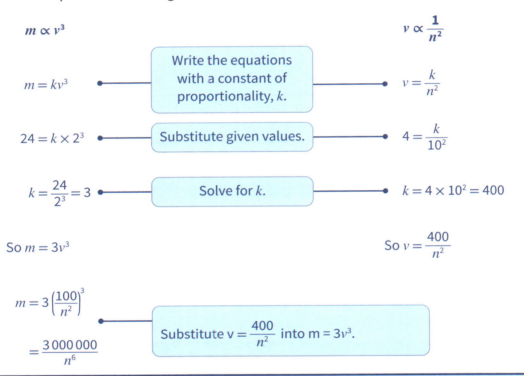

$$m \propto v^3 \qquad\qquad v \propto \frac{1}{n^2}$$

Write the equations with a constant of proportionality, $k$.

$$m = kv^3 \qquad\qquad v = \frac{k}{n^2}$$

Substitute given values.

$$24 = k \times 2^3 \qquad\qquad 4 = \frac{k}{10^2}$$

Solve for $k$.

$$k = \frac{24}{2^3} = 3 \qquad\qquad k = 4 \times 10^2 = 400$$

$$\text{So } m = 3v^3 \qquad\qquad \text{So } v = \frac{400}{n^2}$$

$$m = 3\left(\frac{100}{n^2}\right)^3$$

Substitute $v = \frac{400}{n^2}$ into $m = 3v^3$.

$$= \frac{3\,000\,000}{n^6}$$

**Key terms** — Make sure you can write a definition for these key terms

compound measure    constant of proportionality
direct proportion    inverse proportion
unitary method

## 17 Compound measures and multiplicative reasoning

Learn the answers to the questions below, then cover the answers column with a piece of paper and write as many as you can. Check and repeat.

| | Questions | Answers |
|---|---|---|
| 1 | How do you convert mm into m? | Divide by 1000 (divide by 10 to convert to cm and then by 100 to convert to m). |
| 2 | How do you convert from kg to g? | Multiply by 1000. |
| 3 | How do you convert from ml to l? | Divide by 1000. |
| 4 | How do you convert $cm^2$ into $mm^2$? | Multiply by $10^2 = 100$. |
| 5 | How do you convert $km^2$ into $cm^2$? | Multiply by $100\,000^2 = 10\,000\,000\,000$ (Multiply by $1000^2 = 1\,000\,000$, and then by $100^2 = 10\,000$) |
| 6 | What is the formula for speed? | $\text{speed} = \text{distance} \div \text{time}$ |
| 7 | What is the formula for density? | $\text{density} = \text{mass} \div \text{volume}$ |
| 8 | What is the formula for pressure? | $\text{pressure} = \text{force} \div \text{area}$ |
| 9 | How can you calculate time if you know the distance and speed? | $\text{time} = \text{distance} \div \text{speed}$ |
| 10 | How do you convert km/h to m/s? | Multiply by 1000 and then divide by 60. |
| 11 | What would a graph showing direct proportion look like? | A straight line through the origin. |
| 12 | What is the equation for direct proportion? | $y = kx$ |
| 13 | What is the equation for inverse proportion? | $y = \frac{k}{x}$ |

*Put paper here*

### Previous questions

Now go back and use these questions to check your knowledge of previous topics.

| | Questions | Answers |
|---|---|---|
| 1 | What does it mean to factorise fully? | Make sure you have used the highest common factor, not just a common factor. |
| 2 | What is a variable? | A letter whose value can change. |
| 3 | What is an identity? | An equation that is true for all values. |

*Put paper here*

# Practice

## Exam-style questions

**17.1** Water is poured into an empty tank at a rate of 20 cm³/s.

Work out the time it will take for the tank to contain 2400 cm³ of water.

Give your answer in seconds. **[2 marks]**

**17.2** 5 m³ of wrought iron has mass 38 700 000 g.

Work out the density of the wrought iron.

Give your answer in kg/m³. **[3 marks]**

**17.3** A garden snail travels 5.64 m at a speed of 0.047 km/h.

Work out how long it takes. Give your answer in minutes
and seconds. **[4 marks]**

**EXAM TIP**

Think about units.

**17.4** A 12 acre plot of land in France costs 58 800 euros.

A 15 acre plot of land in Argentina costs 4 520 000 Argentine pesos.

1 euro = 1.15 pounds sterling

1 pound sterling = 70.12 Argentine pesos

Work out which plot of land costs less per acre.

Show your working **[4 marks]**

**17.5** A team of 12 painters can paint a house in 6 days.

**(a)** Calculate how long it would take 18 painters. **[2 marks]**

**(b)** Work out how many painters would be needed to paint the
house in 3 days. **[2 marks]**

**(c)** Sketch a graph to show the relationship between the number
of painters and the time it takes to paint the house. **[2 marks]**

**17.6** $T$ is inversely proportional to $W$.

$T = 5$ when $W = 4$

**(a)** Work out an equation connecting $T$ and $W$. **[3 marks]**

**(b)** Find the value of $T$ when $W = 8$ **[2 marks]**

**17.7** The mass of a statue is $M$ kg.

The height of the statue is $H$ metres.

$M$ is directly proportional to the square of $H$.

$M = 500$ when $H = 0.6$

**(a)** Work out an equation connecting $M$ and $H$. **[2 marks]**

**(b)** Work out the mass of a statue of height 1.5 metres. **[2 marks]**

**17.8**  $f$ is inversely proportional to $g$.

$g$ is directly proportional to $h^2$.

Given that $g = 3$ and $h = 0.5$, when $f = 6$, find a formula for $f$ in terms of $h$. **[4 marks]**

**17.9** An object with volume 100 cm³ has mass of 30 kg.

Pat says, 'If the volume was increased by 40 cm³ and the mass was increased by 40 kg, the density would increase by more than 60%.'

Is Pat correct? You must show working to support your answer. **[2 marks]**

> **EXAM TIP**
>
> Remember:
>
> $Density = \dfrac{Mass}{Volume}$

**17.10** $f$ is directly proportional to $\sqrt{g}$.

$f = 2$ when $g = 324$.

$g$ is inversely proportional to $h^2$.

$g = 225$ when $h = 0.2$.

Find a formula for $f$ in terms of $h$.

Give your answer in its simplest form. **[5 marks]**

**17.11** $v$ is inversely proportional to $w^2$.

$v = 2$ when $w = 3x$

Show that $v = 0.72$ when $w = 5x$ **[3 marks]**

**17.12**  A cheetah can travel 1.98 km in 1 minute.

The distance is measured correctly to the nearest 10 m.
The time is measured to the nearest 5 seconds.

By considering bounds, work out the average speed in m/s of the cheetah to a suitable degree of accuracy.

You must show all your working. **[5 marks]**

**Questions referring to previous content**

**17.13** The mass of an egg decreases by 11% when you remove the shell.

An egg without shell has mass 44.5 g.

Calculate the mass of the egg with the shell on. **[3 marks]**

**17.14** Fred and Ted have some money in the ratio 2 : 1

They both pay £9 for lunch.

The ratio of their money is now 5 : 2

How much money did they each have originally? **[4 marks]**

## 18 Polygons, angles, and parallel lines

## Types of angle

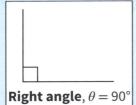
**Right angle**, $\theta = 90°$

**Acute angle**, $0° < \theta \leq 90°$

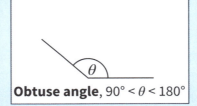
**Obtuse angle**, $90° < \theta < 180°$

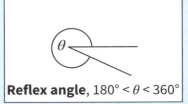
**Reflex angle**, $180° < \theta < 360°$

Two lines are **perpendicular** when the angle between them is exactly 90°

Angles can be measured in a clockwise ↻ or anticlockwise ↺ direction.

## Angle facts

Vertically **opposite angles** are equal.

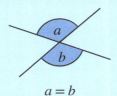

$a = b$

**Parallel lines** are the same distance x from each other. Parallel lines never meet. They are shown on diagrams by arrows on the lines.

Angles $a$ and $b$ are **corresponding angles**. Corresponding angles are always equal.

Angles $c$ and $d$ are **alternate angles**. Alternate angles are always equal.

> **WATCH OUT**
>
> In your exams, make sure you use the correct description (alternate angles, corresponding angles, angles at a point, and so on) or you could miss out on marks.

### Worked example

Work out the size of angles $x$, $y$ and $z$ in the diagram.
Give reasons for your answers.

$x = 75°$ ● ⟶ Vertically opposite angles are equal.

$y = 180° - 75°$ ● ⟶ Angles on a straight line add up to 180°.
$\quad = 105°$

$z = y$ ● ⟶ Corresponding angles are equal.
$\quad = 105°$

## Angles in quadrilaterals

A **quadrilateral** is a 2D shape that has four straight sides.

Angles in a quadrilateral add up to 360°

Square

Rectangle

Kite

Rhombus

Parallelogram

Trapezium

---

**Worked example**

*ABCD* is a parallelogram.

Work out the values of $x$ and $y$.

$\angle DAB = \angle BCD$ — Opposite angles in a parallelogram are equal.

$4x + 15 = 3x + 4y$

$4y - x = 15$ ① — Label this equation ①.

$\angle ABC = 2y$ — Opposite angles in a parallelogram are equal.

So:

$(4x + 15) + 2y + (3x + 4y) + 2y = 360$ — Angles in a parallelogram add to 360°.

$7x + 8y + 15 = 360$

$7x + 8y = 345$ ② — Label this equation ②.

$4y - x = 15$ — Write ① and ② as simultaneous equations.

$7x + 8y = 345$

$4y - x = 15$

$x = 4y - 15$ — Rearrange ① to make $x$ the subject.

$7x + 8y = 345$

$7(4y - 15) + 8y = 345$ — Substitute $x = 4y - 15$ into ②, and solve to find the value of $y$.

$28y - 105 + 8y = 345$

$36y = 450$

$y = 12.5$

$y = 12.5$

$x = 4 \times 12.5 - 15$ — Substitute the value of $y$ into the ① to find the value of $x$.

$x = 35$

# ⚙ Knowledge

## 18 Polygons, angles, and parallel lines

## Angles in polygons

A polygon is a two-dimensional shape with three or more straight edges.

A **regular polygon** (such as an equilateral triangle) has equal sides and equal angles.

An **irregular polygon** has at least one side or angle that is different from the others.

## Interior and exterior angles

An **interior angle** is the angle inside a polygon at a **vertex**.

An **exterior angle** is the angle between a side of the shape and a line extended from the side.

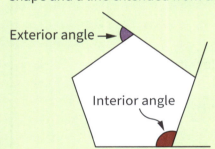

Exterior angle →

Interior angle

sum of interior angles = (number of sides − 2) × 180°

The number of triangles you can split a shape into is always two fewer than the number of sides.

The angles in a triangle add up to 180°.

There are six triangles in an octagon, so

sum of interior angles = 6 × 180°
= 1080°

**Worked example**

A heptagon has interior angles as shown. Work out the size of angle $x$.

sum of interior angles
$= (7 - 2) \times 180°$
$= 5 \times 180°$ ←————————
$= 900°$

$70° + 205° + 45° + 150° + 130° + 110° = 710°$

$x = 900° - 710°$
$= 190°$

Add the six angles you know.

Subtract the sum of the known angles from the sum of all angles.

## Interior and exterior angles

The sum of the exterior angles is the same for any polygon.

sum of exterior angles $= 360°$

### Worked example

The exterior angle of a regular polygon is $15°$.

How many sides does the polygon have?

number of sides exterior angles $= 360°$

number of sides $= \dfrac{360°}{\text{exterior angles}}$ — Rearrange to make 'number of sides' the subject.

number of sides $= \dfrac{360°}{15}$ — Substitute

$= 24$

The shape has 24 exterior angles, so it has 24 sides.

### Worked example

The diagram shows a regular hexagon.

Work out the size of angle $x$.

Find the angle that corresponds to $x$ and draw it on the diagram.

interior angle of a hexagon $= a + 48°$ — Write the values you know in an equation.

sum of interior angles $= (6 - 2) \times 180°$

$= 720°$ — Use the sum of interior angles to find the value of one interior angle.

So, one interior angle $= \dfrac{720°}{6} = 120°$

$a + 48° = 120°$

$a = 120° - 48°$ — Substitute to find value of $a$.

$= 72°$

Corresponding angles are equal, so:

$x = a$ — Work out $x$.

$= 72°$

# ⚙ Knowledge

## 18 Polygons, angles, and parallel lines

### Interior and exterior angles

You can also use the exterior angle to find the interior angle.

**Worked example**

Work out the size of an the interior angle of a regular heptagon.

$$\text{exterior angle} = \frac{360°}{\text{number of sides}}$$

$a = \dfrac{360°}{7}$ ●——— A heptagon has seven sides.

$\quad = 51.4°$

$a + b = 180°$ ●——— $a$ and $b$ lie on a straight line.

$b = 180° - 51.4°$

$\quad = 128.6°$ ●——— Substitute $a = 51.4°$ to find $b$.

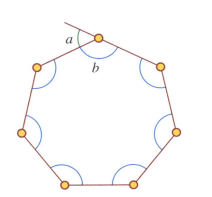

**Worked example**

The diagram shows a regular octagon.

Calculate the value of $x$.

Give a reason for each stage of your working.

sum of interior angles = (number of sides − 2) × 180°

Octagon has 8 sides, so

sum of interior angles = (8 − 2) × 180°

$\qquad\qquad\qquad\qquad = 1080°$

The octagon is regular, so each interior angle is the same. So,

$\dfrac{1080°}{3} = 135°$

$\angle CBA = 135°$

CB and BA are both sides of the regular octagon, so CB = BA.

So CBA is an isosceles triangle.

So $\angle BCA = \angle BAC = x$

$x + x + 135° = 180°$ ●——— Angles in a triangle sum to 180°

$\qquad\quad 2x = 45°$

$\qquad\quad x = 22.5°$

**Key terms**

**Make sure you can write a definition for these key terms**

acute angle   alternate angle   corresponding angle
exterior angle   interior angle   irregular polygon
obtuse angle   parallel   quadrilateral   regular polygon
perpendicular   reflex angle   right angle   transversal   vertex

## 18 Polygons, angles, and parallel lines

Learn the answers to the questions below, then cover the answers column with a piece of paper and write as many as you can. Check and repeat.

| | Questions | Answers |
|---|---|---|
| 1 | What is an acute angle? | An angle that is less than 90°. |
| 2 | What is a reflex angle? | An angle that is greater than 180° and less than 360°. |
| 3 | What is an interior angle? | The angle between two lines on the inside of a shape. |
| 4 | What do angles at a point add up to? | 360° |
| 5 | What do angles on a straight line add up to? | 180° |
| 6 | What do the angles in a triangle sum to? | 180° |
| 7 | What are the properties of a trapezium? | A trapezium is a quadrilateral with one pair of parallel sides. |
| 8 | What is a polygon? | A 2D shape with three or more straight sides. |
| 9 | What is a regular polygon? | A polygon with sides of equal length and angles of equal size. |
| 10 | What is the formula for the sum of interior angles in a polygon? | sum of interior angles = (number of sides – 2) × 180° |
| 11 | What do the exterior angles of any polygon add up to? | 360° |
| 12 | If you are given one exterior angle of a regular polygon, how do you find the number of sides it has? | 360 ÷ size of the exterior angle |

*Put paper here*

## Previous questions

Now go back and use these questions to check your knowledge of previous topics.

| | Questions | Answers |
|---|---|---|
| 1 | What is a mixed number? | A number and a fraction together. |
| 2 | What percentage is equivalent to $\frac{1}{4}$? | 25% |
| 3 | Why is 2 the only even prime number? | All other even numbers are divisible by 2 and therefore have more than two factors. |
| 4 | What does the gradient of the line measure? | How steep it is. |
| 5 | How do you find the next term in a geometric sequence? | You multiply by the common ratio. |

*Put paper here*

# ✏ Practice

## Exam-style questions

**18.1** *ABCG* is a parallelogram.

Work out the size of angle *x*.

Give a reason for each step
of your working.

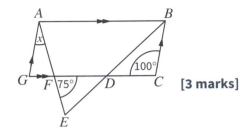

**[3 marks]**

**18.2** The angles in a parallelogram are in the ratio $1 : 3 : 1 : x$

**(a)** Write down the value of *x*. **[1 mark]**

**(b)** Work out the sizes of all four angles. **[2 marks]**

> **EXAM TIP** ◎
>
> The sum of the angles in a
> quadrilateral is 360°.

**18.3** Work out the values of *c* and *d*.
You must show all your working. **[4 marks]**

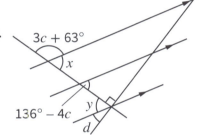

**18.4** A circular pie is cut into four pieces,
as shown. *AC* is the diameter of the pie.

Work out the size of the angle of
the largest slice of pie. Give a reason
for each stage of your working.

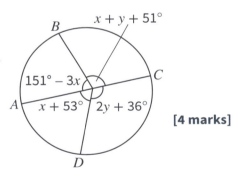

**[4 marks]**

**18.5** Quadrilateral *ABCD* is a rhombus.

Work out the size of angle *x*.

Give full reasons for each step
of your working.

**[3 marks]**

**18.6** In the diagram, angle *BFA* is 44°

Lines AD and BF are parallel.

Work out the size of angle *FDE*.

**[3 marks]**

**18.7** *PQR* and *STU* are congruent equilateral triangles.

*T* is the midpoint of *PR*.

*Q* is the midpoint of *SU*.

Work out the size of angle *t*.

Give reasons for each step of your working.

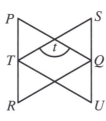

**[3 marks]**

**18.8** Work out the values of *x* and *y* in the kite shown.

**[4 marks]**

**18.9** Part of a regular polygon is shown. One exterior angle has been marked.

Write the name of the polygon

Give a reason for your answer.

**[2 marks]**

**18.10** The sum of the interior angles of a polygon is 1620°.

Work out how many sides the polygon has.

**[2 marks]**

**18.11** The diagram shows a square, two equilateral triangles and two regular octagons.

Work out the size of angle *x*.

**[4 marks]**

> **EXAM TIP**
>
> Consider the rule for the sum of angles round a point.

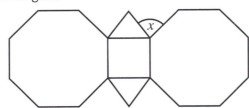

**18.12** Jemima says she has drawn a regular polygon with an interior angle of 80°. Sophia says that it is impossible.

Who is correct? Show working to support your answer.

**[3 marks]**

# Exam-style questions

**18.13** *ABC* and *DEFG* are parallel lines.

Work out the size of angle $x$.

Give a reason for each step of your working.

**18.14** *ABCDEF* is a hexagon.

Angle *BCD* = 2 × Angle *FED*.

Work out the size of angle *FED*.

Show your working.

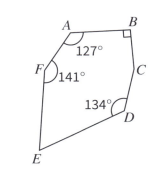

**18.15** *ABCDE* is a pentagon.

The pentagon has one line of symmetry.
*AB* = *AE* and *BC* = *ED*.
Angle *ABC* = 1.5 × Angle *BAE*.

Work out the size of angle BCD.

Show your working.

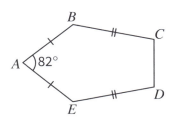

## Questions referring to previous content

**18.16** Solve $4^2 \times 8^2 = \dfrac{1}{2^x}$                                    **[4 marks]**

**18.17** The variable $y$ is indirectly proportional to $\sqrt{2x}$

$y = \dfrac{1}{4}$ when $x = 32$.

Work out the value of $x$ that gives $3\sqrt{2}$        **[5 marks]**

## 19 Area and perimeter (including circles)

## Area

The **area** is the size of the space inside a two-dimensional space.

area of rectangle = base × perpendicular height

$$= bh$$

area of triangle $= \dfrac{(\text{base} \times \text{height})}{2}$

$$= \frac{1}{2} bh$$

area of parallelogram = base × perpendicular height

$$= bh$$

area of trapezium $= \dfrac{1}{2}(a + b)h$

area of circle = π × radius squared

$$= \pi r^2$$

To find the area of a compound shape:

> Split into simpler shapes.

> Find the area of each shape.

> Add up the separate areas.

---

### Worked example

Calculate the area of this shape.

8 cm

7 cm

3 cm

12 cm

> Split the shape into a rectangle and a trapezium.

> Work out $h$: $12 - 8 = 4$ cm

**REVISION TIP**

There will often be more than one way to split the shape up. Can you think of another way to divide this shape to find its area?

total area = area of trapezium + area of rectangle

$$= \frac{1}{2} \times 4(3 + 7) + (8 \times 7)$$

$$= 2(3 + 7) + (8 \times 7)$$

$$= 2(10) + 56$$

$$= 76 \text{ cm}^2$$

> For area, we use units squared because we are multiplying two lengths together.

# ⚙ Knowledge

## 19 Area and perimeter (including circles)

### Area

**Worked example**

The area of the trapezium is 42 cm². Calculate its perimeter.

area of trapezium $= \frac{1}{2}h(a+b)$ ●——— Substitute the values into the area of a trapezium equation.

$42 = \frac{1}{2} \times 6(3+x)$

$42 = 3(3+x)$ ●——— Expand the brackets.

$42 = 9 + 3x$

$33 = 3x$ ●——— Rearrange to make $x$ the subject.

$x = 11$

perimeter $= 11 + 10 + 3 + 6$ ●——— You now have all the values to calculate the perimeter.

$= 30$ cm

### Parts of a circle

Here are the parts of a circle:

**Radius:** The distance of a point on the circle to the centre.

**Diameter:** The distance across the circle through the centre. diameter $= 2 \times$ radius

**Circumference:** The perimeter of a circle.

**Chord:** A line that joins two points on a circle.

**Segment:** The region formed by a chord and the part of the circumference between the two points.

**Tangent:** A line that touches the circumference of a circle in exactly one point, and is perpendicular to the radius at the point of contact.

**Arc:** Part of the circumference of the circle.

**Sector:** Part of the circle formed by two radii and an arc. It is shaped like a slice of cake!

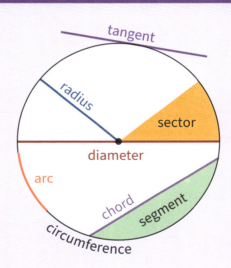

## Circle formulae

### Worked example

Here is a semicircle of radius 4 cm.

(a) Work out the area of the semicircle.
Give your answer correct to 1 d.p.

$$\text{area of circle} = \pi \times 4^2$$
$$= 16\pi \text{ cm}^2$$
$$\text{area of semicircle} = \frac{16\pi}{2}$$
$$= 25.1 \text{ cm}^2 \text{ (to 1 d.p.)}$$

 4 cm

(b) Work out the perimeter of the semicircle.
Give your answer correct to 1 d.p.

$$\text{perimeter of semicircle} = \frac{\text{circumference of circle}}{2} + \text{diameter of circle}$$
$$\text{circumference of circle} = 2 \times \pi \times 4$$
$$= 8\pi \text{ cm}$$
$$\text{circumference of semicircle} = 4\pi$$
$$\text{perimeter of semicircle} = 4\pi + 8$$
$$= 20.6 \text{ cm} \text{ (to 1 d.p.)}$$

> **WATCH OUT**
> Leave your answer in terms of $\pi$ until the very end.

> Identify what you need to calculate.

> Calculate the circumference of the circle, and divide it by 2.

> Add diameter to find the perimeter.

### Worked example

Calculate the area of this compound shape, correct to 1 d.p.

20 cm

6 cm

$$\text{area of rectangle} = 20 \times 6$$
$$= 120 \text{ cm}^2$$
$$\text{radius of semicircle} = \frac{6}{2}$$
$$= 3 \text{ cm}$$
$$\text{area of semicircle} = \frac{\pi \times 3^2}{2}$$
$$= 14.13\ldots \text{ cm}^2$$
$$\text{total area} = 120 + 14.134$$
$$= 134.1 \text{ cm}^2 \text{ (to 1 d.p.)}$$

> The shape is formed of a rectangle and a semicircle. First calculate the area of the rectangle (base × height).

> Now calculate the area of the semicircle.

> **WATCH OUT**
> Remember to calculate the radius first. Do not use the diameter.

# ⚙ Knowledge

## 19 Area and perimeter (including circles)

### Arcs and sectors

For a sector with a radius of $r$ and an angle of $\theta$:

area of sector $= \dfrac{\theta}{360} \times$ area of full circle

$\qquad = \dfrac{\theta}{360} \times \pi r^2$

arc length $= \dfrac{\theta}{360} \times$ circumference of full circle

$\qquad = \dfrac{\theta}{360} \times 2\pi r$

arc

---

**Worked example**

1. Work out the arc length of the sector.

230°

7 cm

> **WATCH OUT** ❗
> Remember to calculate the radius first.

$\quad$ arc length $= \dfrac{\theta}{360°} \times$ circumference of full circle •————— Write out the equation.

$\quad$ arc length $= \dfrac{230}{360°} \times 2 \times \pi \times 7$ •————— Substitute values and calculate.

$\qquad\qquad = 28.1$ cm (to 1 d.p.)

2. Work out the area of the sector. Give your answers to 1 d.p.

$\quad$ sector area $= \dfrac{\theta}{360°} \times \pi r^2$

$\qquad\quad = \dfrac{230}{360°} \times \pi \times 7^2$

$\qquad\quad = 98.3$ cm² (to 1 d.p.)

---

**Worked example**

A circle of radius $r$ has a sector of angle 200° removed.

The perimeter of the remaining sector is 30 cm.

Calculate the value of $r$ to 1 d.p.

perimeter of sector $=$ arc length $+ 2r$

$\quad 30 = \dfrac{200°}{360°} \times 2\pi r + 2r$ •————— Substitute known values into the equation.

$\quad 30 = \dfrac{20}{18} \pi r + 2r$

$\quad 30 = \left(\dfrac{20}{18} \pi + 2\right) r$ •————— Simplify

$\quad r = \dfrac{30}{\dfrac{20}{18} \pi + 2}$ •————— Rearrange to make $r$ the subject.

$\quad r = 5.0$ cm (to 1 d.p.)

> **REVISION TIP**
> A sector can be bigger than a semicircle.

## Arcs and sectors

**Worked example**

$OAB$ is a sector of a circle with centre $O$ and radius 9 m.
The area of the sector is 40.5 m².
Calculate the perimeter of $OAB$.

$$\frac{x}{360°} \times \pi \times 9^2 = 40.5$$

$$x = \frac{40.5 \times 360°}{\pi \times 9^2}$$

$$x = 57.3295°$$

You know the area of the sector, so substitute the values you know into the equation for the area of a sector.

$$\text{arc length} = \frac{57.3295° \dots}{360°} \times 2\pi \times 9$$

$$= 9 \text{ m}$$

Rearrange to make $x$ the subject, and do the calculation to find the value of the angle.

perimeter of sector = radius × 2 + length of arc

$$= 9 \times 2 + 9$$

$$= 27 \text{ m}$$

Calculate the length of the arc.

 **Key terms**

**Make sure you can write a definition for these key terms**

arc   area   chord   circumference   diameter
radius   sector   tangent

## 19 Area and perimeter (including circles)

Learn the answers to the questions below, then cover the answers column with a piece of paper and write as many as you can. Check and repeat.

| | Questions | Answers |
|---|---|---|
| 1 | What is the formula used to find the area of a trapezium? | $\frac{1}{2} \times (a + b)\,h$ |
| 2 | What is the formula used to area of a parallelogram? | length of base $\times$ perpendicular height |
| 3 | What is a compound shape? | A shape made up of two or more simpler shapes. |
| 4 | How can you find the area of a compound shape? | Split it into simpler shapes, find the area of each shape and then add the areas together. |
| 5 | What is a line that touches a circle in exactly one point called? | A tangent. |
| 6 | What is the radius? | The distance from the centre of a circle to the edge. |
| 7 | What is a sector? | A 'slice' of a circle between two radii and an arc. |
| 8 | How do you find the diameter when you know the radius? | Multiply the radius by 2. |
| 9 | What is a chord? | A line that joins two points on a circle. |
| 10 | What is the formula used to find the area of a circle? | $A = \pi r^2$ |
| 11 | What is the formula used to find the circumference of a circle? | $C = \pi d$ or $C = 2\pi r$ |
| 12 | Describe how you find the area of a semi-circle. | Find the area of the circle using $A = \pi r^2$ and divide the answer by 2. |

*Put paper here*

## Previous questions

Now go back and use these questions to check your knowledge of previous topics.

| | Questions | Answers |
|---|---|---|
| 1 | What is rounding? | Writing a number to a given place value. |
| 2 | What is the top number in a fraction called? | The numerator. |
| 3 | What is the general equation of a cubic graph? | $y = ax^3 + bx^2 + cx + d$ |
| 4 | How do you find the next term of an arithmetic sequence? | Add the common difference. |

*Put paper here*

## Exam-style questions

**19.1** The perimeter of a right-angled triangle is 90 cm.

The lengths of its sides are in the ratio 5 : 12 : 13

Work out the area of the triangle. **[3 marks]**

 **19.2** A circle has area $25\pi$ cm$^2$.

Work out the circumference of the circle in terms of $\pi$. **[2 marks]**

**EXAM TIP**
First find the radius of the circle.

**19.3** Work out the area and perimeter of this semicircle.
Give your answers correct to 1 decimal place. **[2 marks]**

9 cm

**19.4** The circular disc shown has been cut into six identical slices.

The radius of the disc is 18 cm.

**(a)** Write the size of angle $x$. **[1 mark]**

**(b)** Work out the area of one sector of the disc.

Give your answer in terms of $\pi$. **[2 marks]**

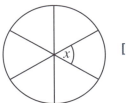

$x$

**19.5** This sculpture is made using three semicircles and is arranged as shown.

Work out the perimeter of the shaded part of the design.

Give your answer to 3 significant figures. **[3 marks]**

19.3 m    4.9 m

**EXAM TIP**
You will need to know the diameters of all 3 circles.

**19.6** A compound shape is made by joining a semicircle on to an isosceles trapezium.

Work out the total area of the shape.

Give your answer to 1 decimal place. **[4 marks]**

11 cm    10 cm

7 cm

**19.7** A trapezium and a triangle have the same area.

The trapezium has parallel sides of length 9.4 cm and $x$ cm, and a height of 12.8 cm.

The triangle has a base of length 12.8 cm and a height of 17.9 cm.

Work out the length $x$. **[3 marks]**

**19.8** The diagram has been created using two identical quarter circles.

Work out the exact area of the shaded region.

**[3 marks]**

> **EXAM TIP**
>
> You will need to find the area of the quarter circles and the square.

**19.9** Here is a sector of a circle.

Work out

**(a)** the area of the sector **[3 marks]**

**(b)** the perimeter of the sector. **[3 marks]**

> **EXAM TIP**
>
> The perimeter of a sector is the sum of the arc length and twice the radius.

**19.10** A pizza has radius of 12 cm.

Jason takes a slice of the pizza.

The remaining pizza has a perimeter of 90 cm.

Work out the angle, $\theta$, of Jason's slice of pizza.

Give your answer to the nearest degree.

**[5 marks]**

**19.11** AOB is a sector of a circle with radius 17 cm.

The area of the sector is 60 cm².

Work out the perimeter of the sector.

Give your answer to 3 significant figures.

**[4 marks]**

**19.12** The diagram shows a rectangle
*ABCD* with length 10 cm.
*AED* and *BEC* are congruent
quarter circles.

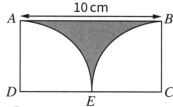

Show that $\dfrac{\text{Area of shaded region}}{\text{Area of rectangle}} = \dfrac{4 - \pi}{4}$. **[4 marks]**

**Questions referring to previous content**

 **19.13** Line $l_1$ intersects the coordinate axes at points *A* and *B* as shown. *B* has
coordinates (0, 5).

Line $l_2$ passes through the points $(a, 7)$ and (7, 16) and is
perpendicular to line $l_1$.

The area of triangle *OAB* is 25 square units.

Work out the value of $a$. **[4 marks]**

 **19.14 (a)** Write 168 as a product of its prime factors. **[3 marks]**

**(b)** When a number, $n$, is cubed, the answer is the same as the
answer to $168 \times 441$.

$441 = 3^2 \times 7^2$

Work out the value of $n$. **[2 marks]**

## 20 Surface area and volume

### 3D shapes

A **three-dimensional (3D) shape** is a solid shape.

3D shapes have **faces**, **edges** and **vertices** (singular: **vertex**).

Faces are the flat surfaces of a 3D shape.

Edges are the lines where two faces meet.

Vertices are the points where three or more edges meet (sometimes called a corner).

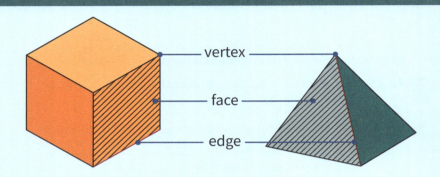

vertex

face

edge

Prisms have a polygon as a base and a constant cross section joined by rectangles. A polygon is any 2D shape with straight sides.

**Cube**

**Cuboid**

**Triangular prism**

cross-section

Note that a **constant** cross section means it is the same from one end to the other. Cubes and cuboids have more than one constant cross section.

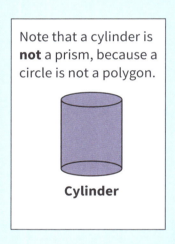

Note that a cylinder is **not** a prism, because a circle is not a polygon.

**Cylinder**

**Sphere**

A **sphere** is a 3D shape with one curved surface. It is the shape of a ball.

**Polyhedron**

A **polyhedron** is a solid shape with many faces. This one is called a dodecahedron; it has ten faces.

**Pyramid**

A **pyramid** is a 3D shape with flat faces. It has a polygon base and triangular faces which meet at a point.

## Volume and surface area formulae

**Volume** (V) is the amount of space inside a 3D shape. It is measured in cubic units (cm³, m³, and so on).

**Surface area** (SA) is the sum of the area of each face. It is measured in square units (cm², m², and so on).

### Cuboid

$V = \text{length} \times \text{width} \times \text{height}$
$= 5 \times 2 \times 3$
$= 30 \text{ cm}^3$

$SA = \text{sum of each face}$
$= 2(5 \times 2 + 5 \times 3 + 2 \times 3)$
$= 62 \text{ cm}^2$

### Prism

$V = \text{area of base} \times \text{length}$

$\text{area of base} = \frac{1}{2} \times l \times w$

$= \frac{1}{2} \times 3 \times 6$
$= 9 \text{ cm}^2$

$V = 9 \times 12$
$= 108 \text{ cm}^3$

$SA = (2 \times \text{triangular face}) +$
$(2 \times \text{sloping face}) + \text{base}$

$\text{area of one triangular face} = \frac{1}{2} \times bh$

$= \frac{1}{2} \times 3 \times 6$
$= 9 \text{ cm}^2$

$\text{area of one sloping face} = 12 \times 6.7$
$= 80.4 \text{ cm}^2$

$\text{area of base} = 12 \times 3$
$= 36 \text{ cm}^2$

$SA = (2 \times 9) + (2 \times 80.4) + 36$
$= 214.8 \text{ cm}^2$

### Cylinder

$V = \text{area of circular face} \times \text{height}$
$= \pi r^2 h$

$V = \pi \times 3^2 \times 8$
$= 72\pi \text{ cm}^3$

$SA = \text{area of curved surface} +$
$2 \times \text{circular face}$
$\text{area of curved surface} = 2\pi rh$
$= 2\pi \times 3 \times 8$
$= 48\pi$

$\text{area of circular face} = \pi r^2$
$= \pi \times 3^2$
$= 9\pi$

$SA = 48\pi + 2 \times 9\pi$
$= 66\pi \text{ cm}^2$

**REVISION TIP**

The area of the curved surface is a rectangle.

Its width wraps around the circumference, so its width is $C = 2\pi r$. This is multiplied by the height to get the area of the rectangle.

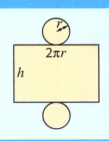

## 20 Surface area and volume

### Volume and surface area formulae

**Sphere**

$V = \dfrac{4}{3}\pi r^3$

$= \dfrac{4}{3}\pi \times 3^3$

$= 36\,\pi\ \text{cm}^3$

$SA = 4\pi r^2$

$= 4\pi \times 3^2$

$= 36\ \text{cm}^2$

**Hemisphere**

$V = \dfrac{1}{2} \times \text{area of sphere}$

$= \dfrac{1}{2} \times \dfrac{4}{3}\pi r^3$

$= \dfrac{2}{3}\pi r^3$

$= \dfrac{2}{3}\pi \times 10^3$

$= \dfrac{2000\pi}{3}\ \text{cm}^3$

SA = area of curved surface + area of base

$= \dfrac{1}{2}$ SA of sphere + area of base

area of curved surface $= \dfrac{1}{2} \times 4\pi r^2$

area of base $= \pi r^2$

area of curved surface $= \dfrac{1}{2} \times 4\pi r^2$

$= 2\pi \times 10^2$

$= 200\pi\ \text{cm}^2$

area of base $= \pi r^2$

$= 100\pi\ \text{cm}^2$

$SA = 100\pi + 200\pi$

$= 300\pi\ \text{cm}^2$

**Cone**

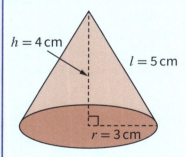

$V = \dfrac{1}{3}\pi r^2 h$

$= \dfrac{1}{3}\pi \times 3^2 \times 4$

$= 12\,\pi\ \text{cm}^3$

SA = area of curved surface + area of base

area of curved surface $= \pi r l$

$= \pi \times 3 \times 5$

$= 15\,\pi\ \text{cm}^2$

area of base $= \pi r^2$

$= 9\pi\ \text{cm}^2$

$SA = 15\pi + 9\pi$

$= 24\pi\ \text{cm}^2$

 **REVISION TIP**

Make sure you use the vertical height, $h$, when calculating the volume.

 **REVISION TIP**

Make sure you use the slant height, $l$, when calculating the volume.

## Volume and surface area formulae

### Pyramid

22.9 cm

23 cm

9 cm

$V = \frac{4}{3}$ area of base × height

area of base = l × w

$= \frac{1}{3}(9 \times 9) \times 23$

$= 621 \text{ cm}^3$

SA = area of triangluar faces + area of base

area of one triangular face $= \frac{1}{2} \times 9 \times 22.9$

$= 103 \text{ cm}^2$

area of base $= 9^2$

$= 81 \text{ cm}^2$

SA $= 4 \times 103 + 81$

$= 493 \text{ cm}^2$

### Worked example

A sphere has surface area $36\pi$ cm².

Work out the volume of the sphere in terms of $\pi$.

$4\pi r^2 = 36\pi$

$\frac{4\pi r^2}{4\pi} = \frac{36\pi}{4\pi}$

$r^2 = \frac{36\pi}{4\pi}$

$r^2 = 9$

$r = \sqrt{9}$

$r = 3$

$V = \frac{4}{3} \times \pi \times 3^3$

$= 36\pi \text{ cm}^3$

Substitute $36\pi$ cm² into the formula for the surface area of a sphere.

Rearrange to make $r$ the subject.

Substitute the value of $r$ into the formula for the volume of a sphere

Leave your answer in terms of $\pi$.

### Worked example

A fence post is the shape of a cone on top of a cylinder.

The height of the cylinder is 10 m.

The radius of the cylinder is 1 m

The height of the cone is 2 m.

Work out the volume of the fence post.

$V_{cylinder}$ = area of circular face × height

$= \pi \times 1^2 \times 10$

$= 10\pi$

$V_{cone} = \frac{1}{3} \times$ area of base × height

$= \frac{1}{3} \times \pi \times 1^2 \times 2$

$= \frac{2}{3}\pi$

total volume $= 10\pi + \frac{2}{3}\pi$

$= 33.5 \text{ m}^3$

Write the formula for the volume of the cylinder. (Remember: area of a circle is $\pi r^2$)

Substitute the known value and calculate.

Write the formula for the volume of the cone.

Substitute the known value and calculate.

Add the two volumes together.

## 20 Surface area and volume

### Volume and surface area formulae

**Worked example**

A cylindrical flask has internal surface area 93 000 mm² and volume 2.2 litres.

1. Write the volume of the flask in cubic millimetres.
   (2.2 litres = 2200 cm³) •——— Convert litres to mm²

   $2200 \times 10^3 = 2\,200\,000$ mm³

2. The inside of the flask is covered in a special coating which costs 0.3p/cm².

   Work out the total cost of the coating.

   $\dfrac{93\,000 \text{ mm}^2}{10^2} = 930$ cm² •——— Convert the surface area to cm³
   Multiply by the cost per cm³

   cost $= 930 \times 0.3$p

   $= 279$p

   $= £2.79$ •——— Convert into £

> **REVISION TIP**
>
> $1 \text{ cm}^3 = 1 \text{ ml}$

**Worked example**

The diagram shows a metal cuboid.

The cuboid is melted and made into smaller cubes.

Each of the smaller cubes has side length 1.5 cm.

1 m, 3 m, 2 m

What is the greatest number of cubes that can be made?

volume of cuboid $= 1 \times 2 \times 3$ •——— Calculate the volume of the cuboid (length × width × height).

$= 6 \ m^3$

$= 6\,000\,000$ cm³

volume of cuboid $= 1.5^3$ •——— Calculate the volume of the smaller cubes (length³).

$= 3.375$ cm³

number of cubes $= \dfrac{\text{volume of cuboid}}{\text{volume of one cube}}$ •——— Work out how many cubes can be made.

$= \dfrac{6\,000\,000}{3.375}$

$= 1\,777\,777.78$

1 777 777 is the greatest number of cubes that can be made.

> **WATCH OUT**
>
> You are looking for the number of *complete* cubes that can be made, so you need to round down not up!

 **Key terms** — Make sure you can write a definition for these key terms.

constant    cuboid    cone    cylinder    edge    face    hemisphere
polyhedron    prism    pyramid    sphere    surface area
three-dimensional (3D) shape    vertex    volume

## 20 Surface area and volume

Learn the answers to the questions below, then cover the answers column with a piece of paper and write as many as you can. Check and repeat.

| Questions | Answers |
|---|---|
| **1** What is a vertex? | The point where two edges meet. |
| **2** What is the same about the cross-section of any prism? | Its cross-section is a polygon. |
| **3** What is a polygon? | Any 2D shape with thee or more straight edges. |
| **4** Why is a cylinder not a prism? | Because a circle is not a polygon. |
| **5** What is a face of a 3D shape? | A flat surface. |
| **6** What is an edge of a 3D shape? | Where two faces meet. |
| **7** What is the surface area of a 3D shape? | The sum of the areas of the surfaces of a 3D shape. |
| **8** What is the volume of a 3D shape? | The space inside a 3D shape. |
| **9** How do you find the volume of a cuboid? | Multiply the length by the with by the height. |
| **10** How do you find the volume of a prism? | Multiply the area of the cross-section by the length. |
| **11** What units do you use for volume? | Cubic units. |
| **12** Describe how you find the volume of a cylinder. | Find the area of the circular face then multiply the answer by the height. |
| **13** What is the formula used to find the volume of a pyramid? | $volume = \frac{1}{3} \times$ area of base $\times$ height |
| **14** How can a net help you to find the surface area of a 3D shape? | You can find the area of each face and add them together. |
| **15** How many faces does a cylinder have? | 2 |

*Put paper here*

## Previous questions

Now go back and use these questions to check your knowledge of previous topics.

| Questions | Answers |
|---|---|
| **1** How do you convert from ml to l? | Divide by 1000. |
| **2** What is an acute angle? | An angle that measures more than 0 and less than 90° |
| **3** What does $a^2 - b^2$ factorise to? | $(a - b)(a + b)$ |
| **4** To simplify a ratio fully, you divide all the numbers in the ratio by what? | The highest common factor of the numbers. |
| **5** What is the sum of the exterior angles of any polygon? | 360° |

*Put paper here*

# Practice

## Exam-style questions

**20.1** An unopened cylindrical can of baked beans has height 11 cm and radius 4 cm.

Work out

(a) the volume of the can in terms of $\pi$ [2 marks]

(b) the surface area of the can to 3 significant figures. [4 marks]

> **EXAM TIP**
> A closed cylinder has three surfaces.

**20.2** The triangular prisms shown have the same volume.

Work out the height $h$.

State your units. [4 marks]

**20.3** The volume of a solid wooden cube is $2\sqrt{2}$ m³.

A cylindrical hole, of radius 0.25 m, is drilled through the length of the cube and its centre.

Work out the total surface area of the six external faces of the resulting solid. Give your answer in the form $a + b\pi$. [3 marks]

> **EXAM TIP**
> Start by finding the side length of the cube.

**20.4** Karen fills two cylindrical jugs of different sizes with water.

volume of the larger jug : volume of the smaller jug is 2 : 1

The total volume of the two jugs is $2400\pi$ cm³.

The larger jug has a radius of 12 cm.

(a) Work out the height of the larger jug to 3 significant figures. [4 marks]

(b) The radius of the smaller jug is equal to its height. Work out its radius of it to 3 significant figures. [3 marks]

**20.5** The Great Pyramid of Giza has a square base of side length 230 m and a height of 147 m.

Work out the volume of the pyramid to 2 significant figures. **[3 marks]**

147 m

230 m

**20.6** A sphere has a surface area of $400\pi$ cm³.

Work out the radius of the sphere. **[2 marks]**

**20.7** A thin hemispherical bowl has a radius of 25 cm.

25 cm

Work out the maximum amount of water the bowl can hold.
Give your answer in millilitres to 3 significant figures. **[3 marks]**

**20.8** A plastic ice cream model is a solid hemisphere on top of a solid cone.

The cone has radius 6 cm and height 15 cm.

The base of the hemisphere is the same size as the base of the cone.

Work out the volume of plastic in the ice cream model. **[3 marks]**

**20.9** A fish tank is in the shape of a cuboid.
The tank has length 1 m, width 60 cm and height 80 cm.

80 cm

60 cm

1 m

**[2 marks]**

Six 1.5-litre jugs of water, are poured into the tank.
The height of the water in the tank is now $h$ cm.

$h$ cm

Work out the value of $h$. **[4 marks]**

**20.10** Greta wants to cover the whole surface area of this closed cylinder in fabric.

The cylinder has diameter 20 cm and height 40 cm.

Greta has 3000 cm² of fabric.

Work out whether or not Greta has enough fabric.

You must show all your all working.

[3 marks]

**EXAM TIP**

The surface area of a closed cylinder is given by the formula $2\pi rh + 2\pi r^2$.

**20.11** An ornament is the shape of a half cylinder on top of a cuboid.

6 cm

18 cm

20 cm

Work out the volume of the ornament. [4 marks]

## Questions referring to previous content

**20.12** At a post office, there are some medium parcels and large parcels. The parcels are to be posted either 1st or 2nd class.

The ratio of medium parcels to large parcels is 3 : 5.

The ratio of 1st class medium parcels to 2nd class medium parcels is 1 : 2.

The ratio of 1st class large parcels to 2nd class large parcels is 3 : 2.

What fraction of the parcels are posted 1st class? [4 marks]

 **20.13** Which circular sector has the greater area?

    **A** a 60° sector of a circle with radius 10 cm

    **B** a 75° sector of a circle with radius 9 cm [3 marks]

## 21 Pythagoras and 2D trigonometry

### Pythagoras' theorem

**Pythagoras' theorem** for **right-angled triangles** is:

$$a^2 + b^2 = c^2$$

Where:
- $c$ is the **hypotenuse** (longest side) of the triangle
- $a$ and $b$ are the other two sides.

You can use Pythagoras' theorem to find missing sides in right-angles triangles.

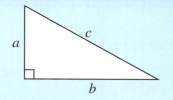

> **REVISION TIP**
>
> You can use Pythagoras' theorem to determine whether a triangle is right-angled.
>
> If three sides of a triangle do not satisfy $a^2 + b^2 = c^2$, then the triangle is not right-angled.

---

**Worked example**

Work out the length of the sides labelled with letters in these triangles.

**1.**
7 cm, $x$, 15 cm

$x^2 = 7^2 + 15^2$ — Substitute the known values into the formula.

$x = \sqrt{7^2 + 15^2}$ — Take the square root of both sides to find $x$.

$= 16.6$ cm (to 1 d.p.)

**2.** 3.5 cm, $y$, 9.8 cm

$9.8^2 = 3.5^2 + y^2$ — Substitute the known values into the formula.

$y^2 = 9.8^2 - 3.5^2$ — Rearrange to make $y$ the subject.

$y = \sqrt{9.8^2 - 3.5^2}$

$= 9.2$ cm (*to* 1 d.p.) — Then solve.

---

**Worked example**

The points $A$ and $B$ have coordinates $(1, 2)$ and $(4, -2)$.

Work out the length of $AB$.

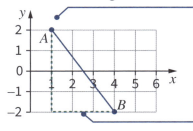

Plot the points $A$ and $B$ on a coordinate grid.

Make a right-angled triangle by drawing a vertical line from $A$ and a horizontal line from $B$.

$(AB)^2 = 3^2 + 4^2 = 25$ — The triangle has a base of 3 and a height of 4.

$AB = \sqrt{25} = 5$

# ⚙ Knowledge

## 21 Pythagoras and 2D trigonometry

## Pythagoras' theorem

**Worked example**

The base of a cone has radius 12 cm. Its volume is $768\pi\,\text{cm}^3$.

Calculate its surface area.

12 cm

**REVISION TIP** ☑

These formulae will be given to you in an exam:

surface area of cone
$= \pi r l + \pi r^2$

volume of cone $= \frac{1}{3}\pi r^2 h$

$$V = \frac{1}{3}\pi \times 12^2 \times h = 768\pi$$

$$48\pi h = 768\pi$$

$$h = 16\,\text{cm}$$

> Height is needed to calculate surface area. So, use the formula for the volume of a cone to find the height.

16 cm

12 cm

> Sketch a right-angled triangle inside the cone.

$$l = \sqrt{16^2 + 12^2}$$

> Use Pythagoras' theorem to find $l$.

$$= 20\,\text{cm}$$

curved surface area $= \pi \times r \times l$

> Find the curved surface area.

$$= \pi \times 12 \times 20$$

$$= 240\pi\,\text{cm}^2$$

area of base $= \pi \times 12^2$

> Find the area of the base.

$$= 144\pi\,\text{cm}^2$$

total surface area $= 240\pi + 144\pi$

> Add together the areas to find the total surface area.

$$= 384\pi$$

$$\approx 1206\,\text{cm}^2$$

## Trigonometry

Use **trigonometry** to find missing side lengths or angles in right-angled triangles.

**Hypotenuse**: The longest side of a right-angled triangle.

**Opposite**: Opposite the angle $\theta$.

**Adjacent**: Next to the angle $\theta$.

Adjacent

$\theta$

Hypotenuse

Opposite

**Formula**

There are three trigonometric ratios that we can use to find missing sides or angles.

$$\sin \theta = \frac{\text{opposite}}{\text{hypotenuse}}$$

$$\cos \theta = \frac{\text{adjacent}}{\text{hypotenuse}}$$

$$\tan \theta = \frac{\text{opposite}}{\text{adjacent}}$$

**REVISION TIP** ☑

Use this mnemonic to remember the trigonometric ratios.

$^S$O$_H$-$^C$A$_H$-$^T$O$_A$

# Solving trigonometry problems

Sometimes you will be given two side lengths and asked to work out an angle.

The method is the same, even though you don't know what the angle is yet.

## Worked example

Work out the value of $x$ in this triangle.

Write down the information you know.

angle = 32°

13 cm side is **adjacent** 32°.

$x$ cm side is **opposite** 32°.

SOH CAH TOA

Underline the side you know and the side you need to find. Use the one that has two underlined letters.

$\tan \theta = \dfrac{\text{opposite}}{\text{adjacent}}$

$\tan 32 = \dfrac{x}{13} \Rightarrow x = 13 \tan 32$

Substitute in the values.

$x = 8.1$ cm (1 d.p.)

Use a calculator.

### WATCH OUT

A lot of students lose marks because they don't know how to use their calculator properly.

Find these buttons:

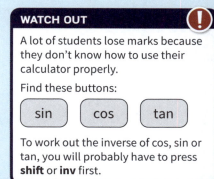

To work out the inverse of cos, sin or tan, you will probably have to press **shift** or **inv** first.

## Worked example

A plank of wood is propped up against a wall. The plank rests 0.5 m from the base of the wall and makes an angle of 70° with the floor.

Work out the length of the plank to the nearest centimetre.

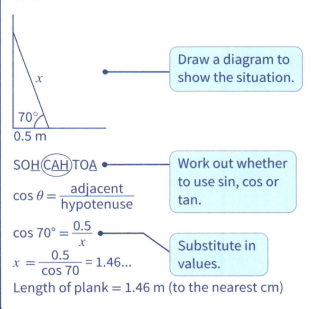

Draw a diagram to show the situation.

SOH CAH TOA

Work out whether to use sin, cos or tan.

$\cos \theta = \dfrac{\text{adjacent}}{\text{hypotenuse}}$

$\cos 70° = \dfrac{0.5}{x}$

$x = \dfrac{0.5}{\cos 70} = 1.46...$

Substitute in values.

Length of plank = 1.46 m (to the nearest cm)

## Worked example

Work out the size of the angle marked $x$ in this triangle.

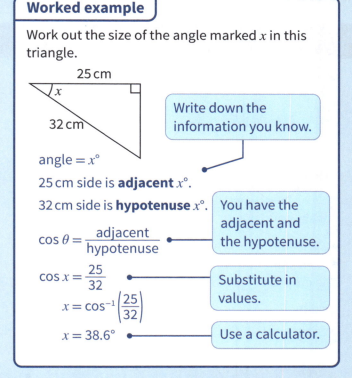

Write down the information you know.

angle = $x°$

25 cm side is **adjacent** $x°$.

32 cm side is **hypotenuse** $x°$.

You have the adjacent and the hypotenuse.

$\cos \theta = \dfrac{\text{adjacent}}{\text{hypotenuse}}$

$\cos x = \dfrac{25}{32}$

Substitute in values.

$x = \cos^{-1}\left(\dfrac{25}{32}\right)$

$x = 38.6°$

Use a calculator.

## 21 Pythagoras and 2D trigonometry

### Trigonometric values

Here are the exact values you need to know (or be able to calculate quickly) in an exam. You can use the triangles or memorise the table.

First, draw a right-angled isosceles triangle with base and height of 1.

Now draw an equilateral triangle with side lengths 2 and split it into two right-angled triangles.

Use Pythagoras' theorem to find these lengths.

| angle, $\theta$ | sin, $\theta$ | cos, $\theta$ | tan, $\theta$ |
|---|---|---|---|
| 0° | 0 | 1 | 0 |
| 30° | $\frac{1}{2}$ | $\frac{\sqrt{3}}{2}$ | $\frac{\sqrt{3}}{3}$ |
| 45° | $\frac{1}{\sqrt{2}} = \frac{\sqrt{2}}{2}$ | $\frac{1}{\sqrt{2}} = \frac{\sqrt{2}}{2}$ | 1 |
| 60° | $\frac{\sqrt{3}}{2}$ | $\frac{1}{2}$ | $\sqrt{3}$ |
| 90° | 1 | 0 | n/a |

### Graphs of trigonometric functions

Learn the shapes and key features of these graphs.

$y = \sin x$

$y = \cos x$

For $\sin x$ and $\cos x$:

- The maximum value is 1
- The minimum value is −1

The graphs of $y = \sin x$ and $y = \cos x$ both repeat every 360°.

$y = \tan x$

$\tan x$:

- does not have a maximum or minimum value.

- can be any value, but is undefined at $x = 90°, 270°, \dots$

The graph of $y = \tan x$ repeats every 180°.

**21**

# Graphs of trigonometric functions

**Worked example**

**(a)** Write the value of sin 60°.

$$\sin 60 = \frac{\sqrt{3}}{2}$$

**(b) (i)** Use your answer to part a) and the graph of $y = \sin x$ to work out the value of sin 120°.

**(ii)** Work out the value of sin 300°.

$$\sin 300° = -\sin 60°$$
$$= -\frac{\sqrt{3}}{2}$$

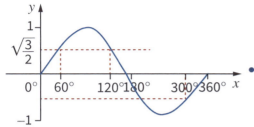

Sketch the sine graph between 0° and 360°.

$$\sin 120° = \frac{\sqrt{3}}{2}$$

**Worked example**

Here is a graph of $y = \sin x$.

On the same axes, sketch the graph of $y = \sin (x - 45°)$ for $0° \le x \le 360°$.

Label the coordinates of the points where $y = \sin (x - 45°)$ intercepts the $x$-axis.

$y = \sin (x - 45°)$ is a translation of $y = \sin x$ by 45° in the direction of the positive $x$-axis. It intercepts the $x$-axis at (45°, 0) and (225°, 0)

**Key terms** Make sure you can write a definition for these key terms

adjacent   cosine (cos)   hypotenuse   opposite
Pythagoras' Theorem   right-angled triangle
sine (sin)   tan   trigonometry

**21 Knowledge**  **165**

## 21 Pythagoras and 2D trigonometry

Learn the answers to the questions below, then cover the answers column with a piece of paper and write as many as you can. Check and repeat.

| | Questions | | Answers |
|---|---|---|---|
| 1 | What kind of triangles can you Pythagoras' theorem be used with? | | Right-angled triangles. |
| 2 | What is the hypotenuse? | Put paper here | The longest side of a right-angled triangle. |
| 3 | What is the value of sin 30°? | | 0.5 |
| 4 | What is the value of tan 45°? | | 1 |
| 5 | What is the formula used to find $\sin \theta$? | Put paper here | $\sin \theta = \dfrac{\text{opposite}}{\text{hypotenuse}}$ |
| 6 | What is the formula used to find $\cos \theta$? | | $\cos \theta = \dfrac{\text{adjacent}}{\text{hypotenuse}}$ |
| 7 | Is : $\dfrac{\text{opposite}}{\text{adjacent}}$ the formula for $\sin \theta$, $\cos \theta$ or $\tan \theta$? | Put paper here | $\tan \theta$ |
| 8 | What is the value of cos 60? | | 0.5 |
| 9 | What is the exact value of sin 60? | | $\dfrac{\sqrt{3}}{2}$ |
| 10 | If $\tan \theta = 5$, how do you find $\theta$? | Put paper here | Use $\tan^{-1}(5)$ |
| 11 | What is the value of cos 90°? | | 0 |
| 12 | When finding the distance between two points on a grid, what theorem can you use? | | Pythagoras' theorem. |

## Previous questions

Now go back and use these questions to check your knowledge of previous topics.

| | Questions | | Answers |
|---|---|---|---|
| 1 | What does LCM stand for? | | Lowest common multiple. |
| 2 | What is the opposite of factorising? | Put paper here | Expanding. |
| 3 | When might you use the quadratic formula? | | To solve a quadratic equation that cannot be factorised. |
| 4 | What do the letters BIDMAS stand for? | Put paper here | Brackets, indices, division, multiplication, addition and subtraction. |

## Exam-style questions

**21.1**  *ABC* is a right-angled triangle.                          **[2 marks]**

Work out the length of *BC*.

Give your answer correct to 3 significant figures.

**21.2**  Point A has coordinates (−1, 3). Point B has coordinates (2, 8).

Work out the length of the line segment AB to 3 significant figures.  **[4 marks]**

**21.3**  The diagram shows the dimensions of a biscuit. The biscuit is in the shape of an isosceles triangle.

Work out the height of the biscuit to 1 decimal place.                          **[3 marks]**

**21.4**  The diagram shows three identical steps. Work out the height *h* of each step to 1 decimal place.                          **[4 marks]**

 **21.5**  A solid cone has height 24 cm and radius 10 cm.

Work out

**(a)**    the volume of the cone                          **[2 marks]**

**(b)**    the surface area of the cone.                          **[4 marks]**

Give your answers in terms of $\pi$.

 **21.6**  Use your calculator to work out $\sqrt{\dfrac{\sin 47° \cos 21°}{\cos 21° - \sin 47°}}$

Give your answer correct to 3 significant figures.                          **[2 marks]**

# Exam-style questions

**21.7** *PQR* is a right-angled triangle.

$\angle QPR = 35°$

$PR = 16$ cm

Work out the value of $x$.

Give your answer correct to 1 decimal place. **[2 marks]**

**21.8** Work out the height $h$ of this trapezium to 3 significant figures. **[3 marks]**

> **EXAM TIP**
> Split the trapezium up into separate shapes.

**21.9 (a)** Sketch the graph of $y = \cos x$ for $-360° \le x \le 360°$. **[2 marks]**

**(b)** Write the coordinates of the first minimum point on the graph where $x > 0$. **[1 mark]**

**21.10** Work out the length of side $b$ in the diagram.
Give your answer to 1 decimal place. **[3 marks]**

**21.11** *MNO* is a right-angled triangle.

$MO = 7$ cm

$NO = 4$ cm **[2 marks]**

Work out the size of the angle marked $x$.

Give your answer correct to 1 decimal place.

**21.12** Triangle *PQR* is an isosceles triangle. Work out angle *PQR* to 1 decimal place.

**[3 marks]**

8 cm

*P*

*Q*          *R*

12 cm

**21.13** *ABC* is a triangle. *AC* = 7 cm, *BC* = 16 cm and angle *BAC* = 90°.

Work out the size of angle *BCA*. **[3 marks]**

 **21.14 (a)** Sketch the graph of $y = \sin x$ for $0° \leq x \leq 450°$. **[2 marks]**

**(b)** The value of sin 45 is $\dfrac{1}{\sqrt{2}}$. Write all the other angles, $\theta°$, for which

$\sin \theta = \dfrac{1}{\sqrt{2}}$ in the interval $0° \leq \theta \leq 450°$. **[2 marks]**

 **21.15** Work out the exact value of *x* in the triangle shown. **[3 marks]**

*x*

$(2 + \sqrt{2})$ cm

$(1 + \sqrt{2})$ cm

**21.16 (a)** Sketch the graph of $y = \tan x$ for $-180° \leq x \leq 360°$. **[2 marks]**

**(b)** Write the value of tan 30°. **[1 mark]**

**(c)** Use the graph to work out the value of

**(i)** tan 210° **[1 mark]**

**(ii)** tan (−30°). **[1 mark]**

**21.17** Sketch the graph of $y = \sin (x + 90)°$ for $0° \leq x \leq 360°$. **[2 marks]**

*y*

1

−180°   −90°   0      90°    180°   *x*

−1

**21.18** Sketch the graph of $y = -\cos x$ for $-180° \le x \le 180°$. **[2 marks]**

**21.19** A squared-based pyramid has a base
length 12 cm and a height of 8 cm.
$F$ is the mid-point of $CB$.

    **(a)** Work out the length $EF$.

    **(b)** Work out the size of angle $ECF$.
        Give your answer to 1
        decimal place. **[3 marks]**

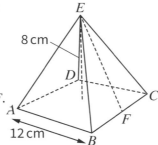

**21.20** $ABC$ is a right-angled triangle.

    **(a)** Work out the size of angle $ACB$.
        Give your answer
        to 1 decimal place. **[2 marks]**

    **(b)** The length of the side $BC$
        is increased. The length of side
        $AB$ remains the same.

        Will the value of tan $ACB$ increase
        or decrease?

        Show working to support your answer. **[1 mark]**

**Questions referring to previous content**

**21.21** $x = 0.7\dot{0}\dot{2}$

    Prove algebraically that $x$ can be
    written as $\frac{26}{37}$. **[3 marks]**

**21.22** The volume of a hemisphere is $60.75\pi$ cm³

    Work out its surface area.

    What do you notice? **[5 marks]**

## 22 Similarity and congruence

## Congruent shapes

**Congruent shapes** are exactly the same size and shape.
- Each pair of corresponding angles are equal.
- Each pair of corresponding sides are equal.
- The shapes may be reflected, rotated or translated, but the triangles are still congruent.

**Corresponding sides** are sides connecting the same angles on two shapes.

There are four conditions that mean two triangles are congruent.

The **three sides** are equal.

**SSS** (side–side -side)

**Two sides** are equal and the **angle** between them are equal.

**SAS** (side – angle – side)

**Two angles** are equal, and one **side** are equal

**ASA** (angle – side – angle)

Both have a **right angle**; and the **hypotenuse** and one other **side** are equal.

**RHS** (right angle, hypotenuse, side)

---

**Worked example**

Explain why the two triangles are congruent.

They are congruent because they satisfy the condition SAS.

$$\text{side } 1 = 4 \text{ cm}$$
$$\text{angle between two sides} = 95°$$
$$\text{side } 2 = 5 \text{ cm}$$

---

**Worked example**

Prove *ABX* and *CDX* are congruent.

>  **REVISION TIP**
>
> There are other ways you could prove that the triangles are congruent – this is just one!

angle *DCB* = angle *ABC* •——— These are pairs of alternate angles.

angle *CDA* = angle *BAD*

    *CD* = *AB* •——— Find the equal sides.

*ABX* and *CDX* are congruent because they satisfy the condition ASA. •——— State the condition the triangles satisfy.

## 22 Similarity and congruence

## Congruent shapes

**WATCH OUT**

If only the three angles are equal (AAA), but not the side lengths, the triangles are **similar**, not congruent.

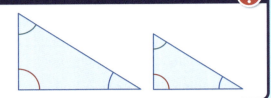

**Worked example**

$A$, $B$ and $C$ are three points on the circumference of a circle.

*angle $ABC$ = angle $ACB$*

$TB$ and $TC$ are tangents to the circle.

Prove that triangles $ATB$ and $ATC$ are congruent.

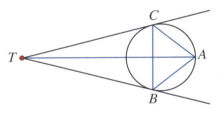

Annotate the image by drawing in lines to make a triangle, and marking any angles you know are the same.

$TB = TC$ (tangents from the same point are equal)

Find the angles that are the same.

$AB = AC$

(angles $ABC$ and $ACB$ are equal, so triangle $ABC$ is isosceles).

$AT$ is a common side, so triangles $ATB$ and $ATC$ are congruent due to SSS.

State the condition the triangles satisfy.

## Similar shapes

**Similar shapes** have exactly the same shape, but different sizes. One shape is an **enlargement** of the other.

The **scale factor** is the number you multiply side lengths of one shape by to get the side lengths of another shape.

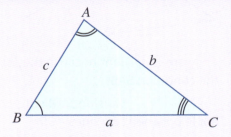

$$\text{scale factor} = \frac{\text{new length}}{\text{original length}}$$

Each pair of corresponding angles are equal

Corresponding sides are in the same proportion:

$A = A_1$, $B = B_1$, $C = C_1$

multiply every side in 1st triangle by 2 to get the lengths in 2nd triangle:

$a = 2 \times a_1$

$b = 2 \times b_1$

$c = 2 \times c_1$

## Similar shapes

### Worked example

Triangles *ABC* and *DEF* are similar.

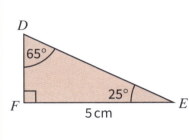

A
3 cm
C  2 cm  B

Not drawn accurately

D
65°
25°
F  5 cm  E

**1.** Calculate the length of *DE*.

|  | triangle *ABC* | triangle *DEF* |
|---|---|---|
| base | 2 | 5 |
| side | 3 | ? |

scale factor $= \dfrac{5}{2}$

length of $DE = 3 \times \dfrac{5}{2}$

$= 7.5$

**2.** Write the size of angle *CAB*.

$\angle CAB = \angle FDE$

$\angle CAB = 65°$

> Make a table to show the corresponding values.

> Use the side that you know on each triangle to calculate the scale factor.

> Multiply the side corresponding to *DE* by the scale factor.

> Find the corresponding angles.

### REVISION TIP

*scale factor for area = (scale factor for length)²*
*scale factor for volume = (scale factor for length)³*

### Worked example

The rectangles are similar.
Find the length of side *x*.

8 cm
12 cm

Not drawn accurately

600 cm²
*x*

$8 \times 12 = 96 \text{ cm}^2$

$\dfrac{600}{96} = 6.25 \text{ cm}^2$

$\sqrt{6.25} = 2.5$

$12 \text{ cm} \times 2.5 = 30 \text{ cm}$

> The area of the second shape has been given, so calculate the area of the first shape.

> Find the scale factor for the area. Square root the area scale factor to find the scale factor for length.

> Multiply the corresponding side length by the scale factor to find the length of side *x*.

A **frustrum** is made by removing a small cone from the top of a larger cone.

### LINK

To remind yourself about finding the volume of a cone, look back at Chapter 20.

### REVISION TIP

For a cone with radius *r*, height *h* and slope height *l*:

volume of cone $= \dfrac{1}{3}\pi r^2 h$

surface area of cone $= \pi rl + \pi r^2$

These formulae will be given to you in an exam.

## 22 Similarity and congruence

### Similar shapes

**Worked example**

Calculate the volume of the frustrum in terms of $\pi$.

30 cm

10 cm

18 cm

The white cone that is removed from the top is similar to the whole cone.
Work out the scale factor for the length.

$$\text{scale factor} = \frac{10}{30}$$

$$= \frac{1}{3}$$

Work out the radius of the larger cone.

$$\text{radius of larger cone} = \frac{18}{2}$$

$$= 9 \text{ cm}$$

use the scale factor to calculate the radius of the smaller cone.

$$\text{radius of smaller cone} = 9 \times \frac{1}{3}$$

$$= 3 \text{ cm}$$

Work out the volume of the larger cone in terms of $\pi$.

$$\text{volume of larger cone} = \frac{1}{3} \pi \times 9^2 \times 30$$

$$= 810\pi \text{ cm}^3$$

Work out the volume of the smaller cone in terms of $\pi$.

$$\text{volume of smaller cone} = \frac{1}{3} \pi \times 3^2 \times 10$$

$$= 30\pi \text{ cm}^3$$

$$\text{volume of frustrum} = 810\pi - 30\pi$$

$$= 780\pi \text{ cm}^3$$

volume of frustrum =
volume of larger cone
− volume of smaller cone

**Worked example**

Here are two similar pyramids.

The surface area of pyramid $A$ is 452 cm²

The surface area of pyramid $B$ is 153 cm²

The volume of Pyramid $B$ is 522 cm³.

Work out the ratio

volume of pyramid $A$ : volume of pyramid $B$.

Pyramid $A$

Pyramid $B$

$$\frac{425}{153} = \frac{25}{9}$$

Work out the scale factor for the surface area.

$$\sqrt{\frac{25}{9}} = \frac{5}{3}$$

Take the square root to work out the scale factor for the length.

$$\left(\frac{5}{3}\right)^3 = \frac{125}{27}$$

Cube your answer to work out the scale factor for the volume.

$$124 : 27$$

Write the ratio.

**Key terms** Make sure you can write a definition for these key terms.

congruent shapes    corresponding sides    enlargement
frustrum    scale factor    similar shapes

## 22 Similarity and congruence

Learn the answers to the questions below, then cover the answers column with a piece of paper and write as many as you can. Check and repeat.

| Questions | Answers |
|---|---|
| **1** What does congruent mean? | Each pair of corresponding angles are equal. Each pair of corresponding sides are equal. |
| **2** What does SSS mean? | Side-Side-Side: All corresponding sides are the same length. |
| **3** What does ASA mean | Angle-Side-Angle: Two pairs of corresponding sides are the same length, and the angle between them is the same size. |
| **4** What does RHS mean? | Right angle-Hypotenuse-Side: Both are right angled triangles. Hypotenuses are the same length and one other side is the same length in both triangles. |
| **5** What does it mean for two shapes to be similar? | They are the same shape, but not the same size. |
| **6** What makes two or more shapes similar? | Each pair of corresponding angles are equal. Corresponding sides are in the same proportion. |
| **7** How do you calculate the scale factor? | $\text{scale factor} = \dfrac{\text{new length}}{\text{original length}}$ |
| **8** Are an object and its enlargement congruent or similar shapes? | They are similar shapes. |
| **9** When using the SAS condition for congruence, which angles need to be equal? | The angle between the two pairs of equal sides. |
| **10** Are all equilateral triangles similar? How do you know? | Yes, as all pairs of corresponding angles are 60°. |

*Put paper here*

## Previous questions

Now go back and use these questions to check your knowledge of previous topics.

| Questions | Answers |
|---|---|
| **1** How do you convert from cm² to mm²? | Multiply by 100. |
| **2** How do you multiply fractions? | Multiply the numerators and multiply the denominators. |
| **3** What is an algebraic expression? | A collection of letters and numbers with no equals sign. |
| **4** What is a quadratic equation? | An equation where the highest power of the variable is 2. |

*Put paper here*

# Practice

## Exam-style questions

**22.1** The two triangles *ABC* and *FED* are congruent.

   **(a)** Write the length of *DF*. **[1 mark]**

   **(b)** Write the size of angle *CAB*. **[1 mark]**

**22.2** *AB* and *DE* are parallel.

*AB = DE*

Prove that triangles *CBA* and *CDE* are congruent. **[4 marks]**

> **EXAM TIP**
> Remember angle rules for parallel lines.

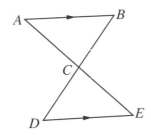

**22.3** *ABCD* is a parallelogram.

*JC* is parallel to *AG*

*BK* is parallel to *ED*

Prove that triangles *ABG* and *CDE* are congruent. **[4 marks]**

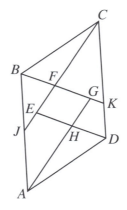

**22.4** Prove that these two triangles are similar. **[2 marks]**

Not drawn accurately

> **EXAM TIP**
> Compare ratios of sides.

**22.5** *EB* is parallel to *DC*.
*ABE* and *ACD* are similar triangles.
*AC* = 11.5 cm, *AB* = 9.2 cm
and *AE* = 8.4 cm.

Work out the length of *ED*. **[2 marks]**

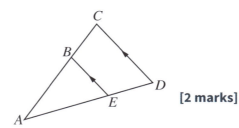

**22.6** *A* and *B* are similar cuboids.
Cuboid *A* has length 10 cm and mass 1.5 kg.

Work out the mass of bar *B*. **[3 marks]**

> **EXAM TIP**
>
> Mass is proportional to volume, so you should start by finding the volume scale factor.

 **22.7** *A* and *B* are similar shapes.
The area of *A* is 12.5 cm$^2$ and
the area of *B* is 50 cm$^2$.
Shape *A* has a base length 4 cm.

Work out the base length of shape *B*. **[3 marks]**

**22.8** Two solids are similar.
The larger solid has surface area 360 cm$^2$
and volume 675 cm$^3$.
The smaller solid has volume 25 cm$^3$.

Work out the surface area of the smaller solid. **[3 marks]**

**22.9** Here are three overlapping trapezia
(large, medium, small) with the same centre.
The two parallel sides of the large
trapezium measure 80 cm and 120 cm.
The medium trapezium is an enlargement
of the large trapezium with scale factor 0.75.
The small trapezium is an enlargement of the medium trapezium with scale
factor 0.5. The area of the shaded region is 4050 cm$^2$.

Work out the height, *h*, of the large trapezium. **[5 marks]**

# Exam-style questions

**22.10** This cork bottle-stopper is in the shape of a frustrum of a cone. The dimensions of the stopper are as shown in the diagram.

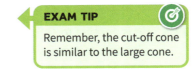
**EXAM TIP**

Remember, the cut-off cone is similar to the large cone.

24 mm

36 mm

18 mm

Work out the volume of the cork.

Give your answer in terms of $\pi$ **[5 marks]**

**22.11** Here are three tetrahedra.

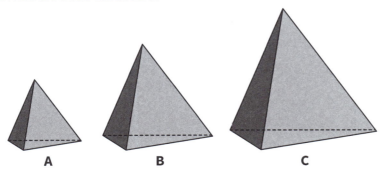

A        B        C

The volume of tetrahedron $B$ is 20% greater than the volume of tetrahedron $A$.

The volume of tetrahedron $B$ is 60% less than than the volume of tetrahedron $C$.

Work out the volume of tetrahedron $A$ as a fraction as the volume of tetrahedron $C$.

Give your fraction in its simplest form. **[3 marks]**

**22.12** Spheres $J$ and $K$ are mathematically similar.
volume of sphere $J$ : volume of sphere $K = 125 : 8$.

The surface area of sphere $K$ is 460 cm³.

Show that the surface area of sphere $J$ is 2875 cm³. **[4 marks]**

**Questions referring to previous content**

**22.13** *ABC* is an isosceles triangle with
*AC* = *BC* = 15 cm and *AB* = 6 cm.

Work out the size of angle *CAB*, giving your
answer to 1 decimal place.                              **[3 marks]**

**22.14** The length of a rectangle is twice the width.

The area of the rectangle is 20 cm².

Work out the length of the rectangle.                  **[2 marks]**

# Knowledge

## 23 Transformations

## Transformations

A **transformation** is something that alters either the size of a shape or its position.

- The original shape is called the **object**.
- The transformed shape is called the **image**.

image

object

## Reflection

When a shape has been reflected, the object and image are **congruent**, but they do not have the same orientation.

### Worked example

Describe the transformation that maps *ABC* to *DEF*.

> Find the two closest corresponding points. Split the distance to find the **line of reflection**. Draw it on the grid.

The transformation is a **reflection** in the line $x = 1$

> A reflection is described by stating the line of symmetry.

**REVISION TIP** ✓

You can use a mirror to help.

## Rotation

When a shape has been reflected, the object and image are **congruent**, but they do not have the same orientation.

### Worked example

Describe the transformation that maps *ABC* to *DEF*.

**REVISION TIP** ✓

You can use tracing paper to help. You can ask for some in your exam!

The transformation is a **rotation** of 90° anticlockwise about the origin.

> A rotation is described by stating:
> - the angle of rotation
> - the direction of rotation
> - the **centre of rotation**.

> Find the corresponding points on the object and image. Find the point that is the same distance from both points and draw it on the grid.

## Transformations

A **translation** can be described using a **column vector**.

- The top number shows the movement in the *x*-axis.
- The bottom number shows the movement in the *y*-axis.

When a shape has been translated, the object and image are **congruent** and they have the same orientation (unlike rotations and reflections).

## Translation

> **Worked example**

1. What is the transformation that maps *ABC* to *DEF*?

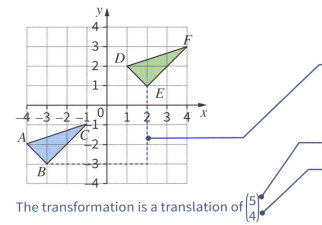

Starting from the object, choose a point and count the horizontal and vertical squares to the corresponding point on the image.

The shape has moved:

5 squares in the positive *x*-direction

4 squares in the positive *y*-direction

The transformation is a translation of $\begin{pmatrix} 5 \\ 4 \end{pmatrix}$

2. What is the transformation that maps *DEF* to *ABC*?

The transformation is a translation of $\begin{pmatrix} -5 \\ -4 \end{pmatrix}$.

To go from the image to the object, the translation is the inverse of the original translation. You can also find this by counting squares on the grid again.

## Enlargement

To enlarge a shape, you need to know:
- the **centre of enlargement**
- the **scale factor**.

If the scale factor is:
- positive, the image will be on the same side of the centre of enlargement as the object, and will have the same orientation as the object
- negative, the image will be on the other side of the centre of enlargement, and will be reversed from the object
- greater than 1, the image will be bigger than the object
- less than 1, the image will be smaller than the object.

For example, an object that is enlarged by $-\frac{3}{4}$ will have an image that is smaller than the object, and will be reversed, on the other side of the centre of orientation.

When a shape has been enlarged, the object and image will be **similar**, not congruent.

 **Key terms**

Make sure you can write a definition for each of these key terms

centre of enlargement    centre of rotation    column vector
congruent    centre of enlargement    image    line of reflection
negative scale factor    object    positive scale factor
reflection    rotation    scale factor    similar
transformation    translation

# Knowledge

## 23 Transformations

## Enlargement

---

**Worked example**

Enlarge triangle $ABC$ by scale factor 2, centre (0, 0).

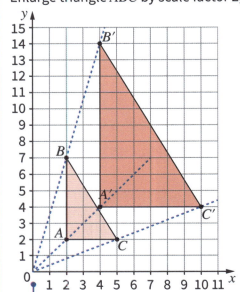

**2** right; **2** up.

$2 \times 2 = \mathbf{4}$; $2 \times 2 = \mathbf{4}$

$(0, 0) \Rightarrow (\mathbf{4}, \mathbf{4})$

$A' = (4, 4)$

$B' = (4, 7)$

$C' = (10, 4)$

Measure the distance of vertex $A$ from the centre of enlargement.

Multiply these numbers by the scale factor.

Add these numbers to the centre of enlargement to find vertex $A'$.

Repeat the process for $B$ and $C$ to plot vertices $B'$ and $C'$.

centre of enlargement

---

**Worked example**

Enlarge the shape by scale factor 0.5, centre (1, 1).

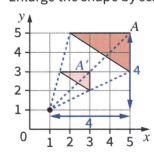

4 right, 4 up

$4 \times 0.5 = 2$

$4 \times 0.5 = 2$

$(1, 1) \rightarrow (3, 3) = A'$

Count the vertical and horizontal distances of vertex $A$ from the centre of enlargement.

Multiply by the scale factor.

Add to centre of enlargement to find $A'$.

---

**Worked example**

Enlarge triangle $A$ by scale factor $-2$, centre (0, 0).

Draw dotted lines from each of the vertices of the triangle through the centre of enlargement.

The distance from the centre of enlargement to each of the vertices is multiplied by 2, but the image is on the opposite side of the centre of enlargement.

---

## 23 Transformations

Learn the answers to the questions below, then cover the answers column with a piece of paper and write as many as you can. Check and repeat.

| Questions | Answers |
|---|---|
| **1** What are the four types of transformation? | Reflection, rotation, translation, enlargement. |
| **2** What do you need to know to draw the reflection of a shape on a graph? | The equation of the line of symmetry. |
| **3** When you reflect a shape, are the object and the image congruent or similar? | Congruent. |
| **4** What three pieces of information do you need to know to describe the rotation of a shape? | The angle, the direction and the centre of rotation. |
| **5** What can you ask for in an exam to help with drawing rotations? | Tracing paper. |
| **6** How do you write a translation mathematically? | Using a column vector. |
| **7** When you translate a shape, are the object and image congruent or similar? | Congruent. |
| **8** What does the top number of a column vector tell you? | How far to move right (if positive) or left (if negative). |
| **9** To enlarge a shape, what two pieces of information do you need? | The scale factor and the centre of enlargement. |
| **10** If you are given the object and the image, how do you find the centre of enlargement? | Draw lines through the corresponding vertices and see where they meet. This point is the centre of enlargement. |

*Put paper here*

## Previous questions

Now go back and use these questions to check your knowledge of previous topics.

| Questions | Answers |
|---|---|
| **1** What is a regular shape? | A shape that has all sides the same length and all angles the same size. |
| **2** What is the formula that is used to calculate the sum of the interior angles in a polygon? | sum of interior angles = (number of sides − 2) × 180° |
| **3** What can you say about the size of two corresponding angles on parallel lines? | They are equal. |
| **4** What is the area of a 2D shape? | A measure of the space inside a 2D shape. |
| **5** How do you find the area of a compound shape? | Split it into simpler shapes, find the area of each shape and then add the areas up. |

*Put paper here*

**23.1**

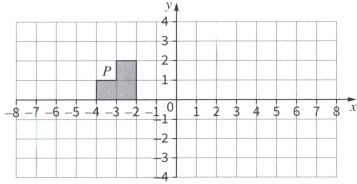

**(a)** Rotate shape $P$ by 180° about the point (1, 0).
Label the new shape $A$. **[1 mark]**

**(b)** Translate the shape $P$ by the vector $\begin{pmatrix} -2 \\ -3 \end{pmatrix}$.
Label the new shape $B$. **[1 mark]**

**23.2** Describe fully the transformation that maps $A$ to $B$. **[2 marks]**

> **EXAM TIP**
>
> If you are unsure, a piece of tracing paper may help you.

**23.3** Enlarge the triangle $P$ by a scale factor of 2 through the point (0, −4). **[3 marks]**

**23.4** **(a)** Enlarge shape $A$ by scale factor $\frac{1}{2}$, centre $(0, 0)$.

Label your image $B$. **[2 marks]**

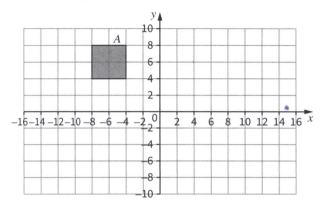

**(b)** Enlarge shape $A$ by scale factor $-1$, centre $(0, 0)$.
Label your image $C$. **[2 marks]**

**(c)** Write two other transformations that map shape $A$ directly
onto shape $C$. **[3 marks]**

**23.5** Rotate shape $P$ 90° clockwise about the point $(-6, 3)$.
Enlarge your image by scale factor $-2$, centre $(-2, 0)$.
Label your final image $Q$. **[4 marks]**

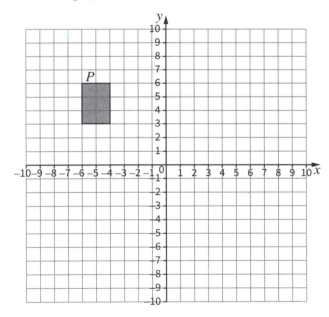

**23.6** Describe fully the single transformation that
maps triangle $A$ onto triangle $B$.

**[3 marks]**

**23.7** Triangle $P$ is reflected in the line $x = 2$ to give triangle $Q$.
Triangle $Q$ is reflected in the line $y = 0$ to give triangle $R$.

Describe the single transformation that maps
triangle $P$ onto triangle $R$. **[3 marks]**

**23.8** Rectangle $D$ is transformed by combining two
single transformations:
rotation 90° anticlockwise, centre (4, −2) followed
by a translation by vector $\begin{pmatrix} 4 \\ 2 \end{pmatrix}$.

One point on rectangle $D$ is invariant under the
combined transformation. Work out the coordinates
of this point. **[2 marks]**

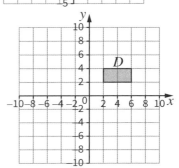

**Questions referring to previous content**

**23.9** A line, $L$, has equation $y = 4 - 3x$

Sajid says that the line $3x + y = 0$ is parallel to $L$.

Is Sajid correct?

Show working to support your answer. **[2 marks]**

**23.10** 100 worms are measured.

The table shows their lengths in cm. **[2 marks]**

| Length ($x$ cm) | Frequency, $f$ | Midpoint | $f \times$ midpoint |
|---|---|---|---|
| $0 < x \le 8$ | 50 | 4 | 200 |
| $8 < x \le 16$ | 30 | 12 | |
| $16 < x \le 24$ | 20 | | 400 |
| Total | 100 | | |

**(a)** Complete the missing values in the table. **[2 marks]**

**(b)** Write the modal class. **[1 mark]**

**(c)** Work out an estimate for the mean length. **[2 marks]**

## 24 Plans, elevations, constructions, bearings

### Plans and elevations

A **plan** is the view from directly above an object.

An **elevation** is the view from the side or the front of an object.

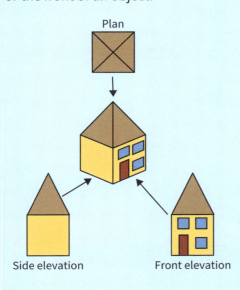

Plan

Side elevation    Front elevation

The 3D solid is made using eight centimetre cubes.

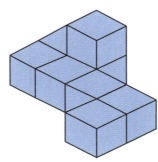

Draw a plan of the solid on a grid.

Start by drawing the outline of the lower layer, then add in lines to show the upper layer.

**REVISION TIP**

The plan can be drawn in any orientation, as long as it's from above.

**Worked example**

The plan view and the front and side elevations of a shape are shown.

| Plan | Front elevation | Side elevation |
|------|-----------------|----------------|

The plan shows the base of the shape is a rectangle of 3 by 2 cubes, with the top left cube raised.

The front and side elevations both shows that the top left part of the plan view is raised by a height of 1 cube.

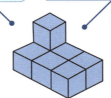

The shape is made up of identical cubes.
Draw a 3D sketch of the shape.

### Constructing triangles

You can use a ruler and a protractor to accurately draw a triangle when you know one of these combinations:

- angle – side – angle (ASA)
- side – angle – side (SAS)
- hypotenuse and one other side of a right-angled triangle (RHS).

## Constructing triangles

**Worked example**

In the triangle $ABC$, $AB = 6$ cm, $BC = 8$ cm and $ABC = 62°$
Draw ther triangle $ABC$.

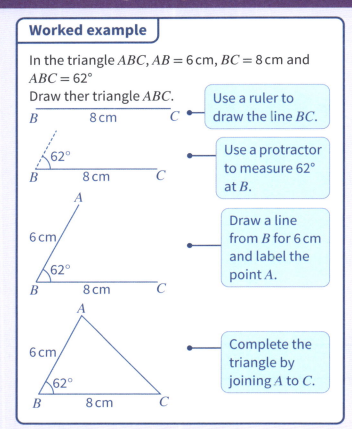

Use a ruler to draw the line $BC$.

Use a protractor to measure 62° at $B$.

Draw a line from $B$ for 6 cm and label the point $A$.

Complete the triangle by joining $A$ to $C$.

**Worked example**

In the triangle $PQR$, $RPQ = 43°$, $RQP = 57°$ and $PQ = 7$ cm.
Draw triangle $PQR$.

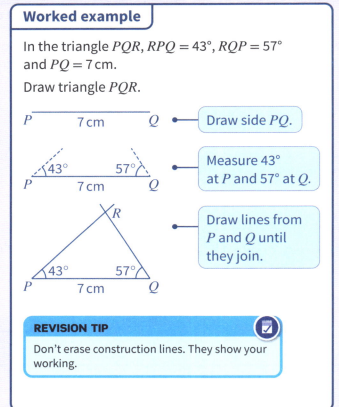

Draw side $PQ$.

Measure 43° at $P$ and 57° at $Q$.

Draw lines from $P$ and $Q$ until they join.

> **REVISION TIP** 🗒
>
> Don't erase construction lines. They show your working.

You can also use a ruler and a **pair of compasses** to **construct** a triangle if you know all three side lengths.

**Worked example**

Construct a triangle with side lengths 7 cm, 4 cm and 5 cm.

Use a ruler to draw a horizontal 7 cm line, $AB$.

Set compasses to 5 cm. Put compasses point at $A$ and draw an arc.

Set compasses to 4 cm. Put compasses point at $B$ and draw an arc.

Draw lines from $A$ and $B$ to where the arcs cross.

# Constructing bisectors

**Perpendicular** lines are at 90° to each other (at right angles).

A perpendicular **bisector** cuts a line exactly in half at right angles. It can therefore be used to find the midpoint of a line.

The perpendicular distance is the shortest distance from a point to a line.

Perpendicular bisector

## Worked example

Construct the perpendicular from the dot though the line.

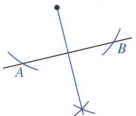

Put compasses on the dot. Draw two arcs on the line. Label the intersects $A$ and $B$.

Put compasses at $A$. Draw an **arc** below the line.

Put compasses at $B$. Draw an arc below the line.

Draw the line from the dot through where the arcs meet.

Angles can also be bisected. A bisector also cuts an angle exactly in half.

## Worked example

Use a ruler and compasses to bisect the angle.

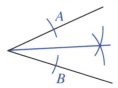

Put compasses on the vertex. Draw an arc on each line. Label the intersects $A$ and $B$.

Put compasses at $A$. Draw an arc to the right of the angle.

Put compasses at $B$. Draw an arc to the right of the angle.

Draw the line from the angle through where the arcs meet.

## Worked example

Construct an angle of 135°

If the angle from the horizontal to a line is 135°, clockwise, then the angle from the horizontal to the same line, anticlockwise, is $180 - 135 = 45°$

An angle of 45° is the bisector of a right angle. So:

Construct the perpendicular bisector (see **blue** arcs and line) .

Construct the angle bisector. (see **red dotted** arcs and line).

**WATCH OUT**

Remember not to delete your construction lines - draw them clearly.

## Loci

**Locus** (plural **loci**) are points that are all the same distance from a point or line.

The **locus of points** a fixed distance from a point is a circle.

All the points that are 2 cm from $P$ lie on the red circle.

The locus of points a fixed distance from a line.

All the points that are 3 cm from $l_1$ lie on the red line, $l_2$.

The locus of points **equidistant** from two points is a perpendicular bisector of those two points.

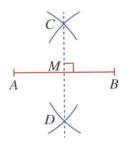

All the points that are equidistant from both $A$ and $B$ lie on the line $CD$.

The locus of points equidistant from two points is a perpendicular bisector of those two points.

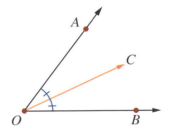

All the points that are **equidistant** from both $OA$ and $OB$ lie on the line $OC$.

---

**Worked example**

A runway is 3 km long. A fence surrounds the runway at a distance of 1.2 km the whole way around.

1. Construct a **scale drawing** of the runway, modelled as a line $AB$, and the fence, using a scale of 1 : 60 000

$$3 \text{ km} = 300\,000 \text{ cm}$$
$$1.2 \text{ km} = 120\,000 \text{ cm}$$ — Convert the distances from km to cm.

length of $AB$ in drawing
$$= \frac{300\,000}{60\,000}$$
$$= 5 \text{ cm}$$ — Divide the lengths by the scale given to find how long the distances should be.

distance to fence in drawing
$$= \frac{120\,000}{60\,000}$$
$$= 2 \text{ cm}$$

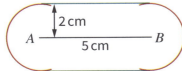

Draw a 5 cm line for the runway, then draw parallel lines 2 cm away and a semicircle of radius 2 cm at each end for the fence.

## Loci

**Worked example**

2. Shade the region where the points inside the fence are closer to *B* than to *A*.

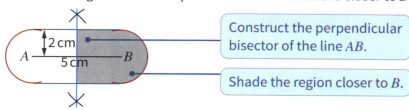

Construct the perpendicular bisector of the line *AB*.

Shade the region closer to *B*.

## Bearings

A **compass** (not a pair of compasses!) shows the direction of magnetic north and the directions east, south and west.

A **bearing** indicates the direction of travel.

Bearings are angles that are measured clockwise from north.

Bearings always have three figures; for example, a bearing of 3° is written as 003°.

North has a bearing of 000°

West has a bearing of 270°

East has a bearing of 090°

South has a bearing of 180°

**Worked example**

A boat sails south-east from a rock.

1. Work out the bearing of the boat from the rock.

South-east is half-way between east (bearing of 090°) and south (bearing of 180°)

The bearing of the boat from the rock is 90° + 45° = 135°

2. Work out the bearing of the rock from the boat.

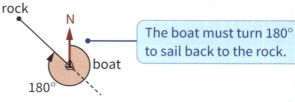

The boat must turn 180° to sail back to the rock.

The bearing of the rock from the boat is 135° + 180° = 315°

**REVISION TIP**

Always draw a diagram!

**LINK**

To remind yourself about measuring angles in degrees look back at Chapter 19.

**Key terms**

Make sure you can write a definition for these key terms

arc    bearing    bisector    compass
construct    due    (north, south east, west)
equidistant    front elevation    locus
pair of compasses    perpendicular    plan
scale drawing    side elevation

# Retrieval

## 24 Plans, elevations, constructions, bearings

Learn the answers to the questions below, then cover the answers column with a piece of paper and write as many as you can. Check and repeat.

| | Questions | | Answers |
|---|---|---|---|
| 1 | What is the first step for drawing a perpendicular bisector of a line segment? | Put paper here | Put the point of the compasses at one end of the line. Open the compasses to more than half the length of the line, and draw arcs above and below the line. |
| 2 | What are perpendicular lines? | | Lines that cross at 90°. |
| 3 | What does bisect mean? | | To cut in half. |
| 4 | What is the perpendicular bisector of an angle? | | A line that cuts an angle in half. |
| 5 | What shape is the locus of points that are a fixed distance from a point? | Put paper here | A circle. |
| 6 | The locus of points a fixed distance from a line form what? | | A line parallel to the original line. |
| 7 | What direction are bearings measured in? | Put paper here | Clockwise from north. |
| 8 | How many figures do bearings have? | | Three. |
| 9 | What is a plan view? | | The view from directly above. |
| 10 | What is a side elevation? | | The view from the side. |

## Previous questions

Now go back and use these questions to check your knowledge of previous topics.

| | Questions | | Answers |
|---|---|---|---|
| 1 | What is a prism? | Put paper here | A 3D shape with a polygon as a constant cross-section. |
| 2 | What is the perimeter of a 2D shape? | | The distance around the edges of a 2D shape. |
| 3 | What is a vertex? | | A point on a 2D or 3D shape where the edges meet. |
| 4 | What is an acute angle? | | An angle that measures less than 90°. |
| 5 | What is the sum of the exterior angles of any polygon? | | 360° |

## Exam-style questions

**24.1** Here are the front and side elevations of a solid shape.

    **(a)** Draw the plan of the solid shape on squared paper. **[1 mark]**

    **(b)** Draw a 3D sketch of the solid shape. **[2 marks]**

**Front elevation**      **Side elevation**

    **(c)** Write the number of the solid's

        **(i)** vertices **[1 mark]**

        **(ii)** edges **[1 mark]**

        **(iii)** faces. **[1 mark]**

**24.2** This solid shape is made from six identical cubes.

Draw the side and front elevations and the plan of the shape on squared paper.

Side elevation

Front elevation

**24.3** **(a)** Only one of these triangles can be constructed. Which one? **[1 mark]**

        **A:** A triangle with sides 7 cm, 8 cm and 16 cm.
        **B:** A triangle with sides 3 cm, 4 cm and 6 cm.
        **C:** A triangle with sides 9 cm, 12 cm and 22 cm.
        **D:** A triangle with sides 15 cm, 30 cm and 60 cm.

    **(b)** Using a ruler and a pair of compasses, construct the triangle you have chosen in part **a**. **[3 marks]**

**24.4** Construct triangle $DEF$ where $DE = 66$ mm, $EF = \frac{2}{3} DE$ and $FD = \frac{3}{4} EF$. **[3 marks]**

**24.5** Using a ruler and compasses, construct the line perpendicular to the line segment $FH$ that passes through $G$. Show all your construction lines **[3 marks]**

> **EXAM TIP**
>
> Start by drawing arcs either side of $G$.

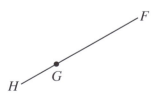

**24.6** Using a ruler and compasses, construct an angle of $45°$. Show your construction lines. **[3 marks]**

# Exam-style questions

**24.7** Using ruler and compasses, show the region inside the grid that is:
less than 3 cm from $B$
nearer to $B$ than to $A$.
Label the region $R$.
Show all your construction lines.

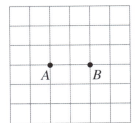

**[4 marks]**

**24.8** $CDE$ represents the plan of a field. Milly farms the area that is closer to $CD$ than to $CE$ and is also less than 500 m from $C$.

Copy the diagram and using ruler and compasses, show the region that Milly farms.

Label the region $R$.

Show all your construction lines.

**[4 marks]**

**24.9** Here is a plan of a T-shaped go-kart track.

There is a barrier all around the track, exactly 2.5 m from the track. Using a scale of 1:125, draw a plan of the barrier.

**[3 marks]**

**24.10** A speedboat ($S$) is due east of a dinghy ($D$).

The bearing of a hovercraft ($H$) from the dinghy ($D$) is 080°.

The bearing of the same hovercraft from the speedboat is 280°.

On the diagram, draw accurately the position of the hovercraft. Label it $H$.

**[3 marks]**

**24.11** Point $B$ is 8 km due north of point $A$.

Point $C$ is 11 km due east of point $A$.

Work out the bearing of $B$ from $C$ to the nearest degree.

**[4 marks]**

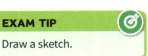

EXAM TIP

Draw a sketch.

**24.12** A buoy is 30 m away from a look-out
post on a bearing of 045°.
A swimmer is 40 m from the same
look-out post on a bearing of 315°.

Work out the distance of the swimmer
from the buoy.

**[4 marks]**

North

 **24.13** The diagram shows the positions of
3 towns, $A$, $B$ and $C$.
The bearing of $B$ from $A$ is 034° $AB = BC$.

Work out the bearing of $A$ from $C$.

**[3 marks]**

### Questions referring to previous content

**24.14** The length, $p$ m, of a football pitch is given as 110 m.
Write the error interval for $p$ if this value is rounded to

**(d)** the nearest 10 metres:   $\leq p <$    **[2 marks]**

**(e)** the nearest 5 metres:   $\leq p$    **[2 marks]**

**24.15** A number, $x$, is given rounded to a particular degree of accuracy.
Write the error interval for $x$ in each case.

**(a)** $x = 4.67$ to 2 decimal places:   $\leq x <$    **[2 marks]**

**(b)** $x = 5000$ to 1 significant figure:   $\leq x$    **[2 marks]**

## 25 Trigonometry in 3D, sine and cosine rules

### Pythagoras' theorem in 3D

You can use **Pythagoras' theorem** twice to find the longest diagonal in a cuboid.

**Worked example**

Work out the length $BG$ in this cuboid.

Sketch triangle $FGH$.

$$(FG)^2 = (GH)^2 + (FH)^2$$
$$(FG)^2 = 5^2 + 2^2$$
$$FG = \sqrt{29} \text{ cm}$$

Use Pythagoras' theorem to work out the length of $FG$.

**REVISION TIP**

You can also use a single step to solve this problem.

$$a^2 + b^2 + c^2 = d^2$$

Sketch triangle $BGF$.

$$(BG)^2 = (BF)^2 + (FG)^2$$
$$(BG)^2 = 3^2 + \sqrt{29}$$
$$BG = \sqrt{38} \text{ cm}$$
$$= 6.16 \text{ cm (3 s.f.)}$$

Use Pythagoras' theorem to work out the length of $BG$.

### Pythagoras and trigonometry in 3D

**Worked example**

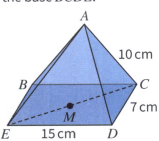

$ABCDE$ is a pyramid with a horizontal rectangular base. $A$ is vertically above $M$, the midpoint of $EC$. Work out the size of the angle between $AC$ and the base $BCDE$.

Sketch $\triangle CDE$.

$$EC^2 = 15^2 + 7^2$$
$$EC = \sqrt{274}$$
$$CM = \tfrac{1}{2}\sqrt{274}$$

Use Pythagoras' theorem to work out $EC$.

Halve the length of $EC$ to find $CM$.

Sketch $\triangle ACM$. Use trigonometry to find angle $ACM$.

$$\cos x = \frac{\sqrt{274}}{20}$$
$$x = \cos^{-1}\frac{\sqrt{274}}{20}$$
$$= 34.1°$$

The angle between $AC$ and the base is 34.1° to 1 d.p.

**25**

# The sine and cosine rules

The **sine rule** and **cosine rule** work for all triangles, not just right-angled ones.

## sine rule

To find an angle use

$$\frac{\sin A}{a} = \frac{\sin B}{b} = \frac{\sin C}{c}$$

To find a side use

$$\frac{a}{\sin A} = \frac{b}{\sin B} = \frac{c}{\sin C}$$

## cosine rule

To find an angle use

$$\cos A = \frac{b^2 + c^2 - a^2}{2bc}$$

To find a side use

$$a^2 = b^2 + c^2 - 2bc \cos A$$

> **REVISION TIP**
>
> Opposite sides and angles have the same letter.
>
> You can decide which angle is $A$, $B$ and $C$.

In some questions you will need to use both trigonometry of right-angled triangles, and the sine or cosine rule.

---

**Worked example**

Work out the size of angle $x$.

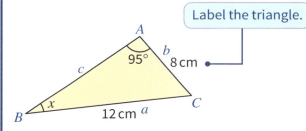

Label the triangle.

$$\frac{\sin A}{a} = \frac{\sin B}{b}$$

$$\frac{\sin x}{8} = \frac{\sin 95°}{12}$$

$$\sin x = \frac{8 \sin 95°}{12}$$

$$= 0.664...$$

$$x = \sin^{-1}(0.664...)$$

$$= 41.6° \text{ (to 1 d.p.)}$$

You know two sides and their corresponding angles, so use the sine rule.

---

**Worked example**

Work out the length of $AB$.

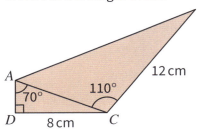

Use the cosine rule to find side $c$.

$$c^2 = a^2 + b^2 - 2ab \cos C$$

$$= 12^2 + (8.51...)^2 - 2 \times 12 \times (8.51...) \cos 110°$$

$$= 286.36...$$

$$c = \sqrt{286.36...}$$

$$= 16.9 \text{ cm (to 1 d.p.)}$$

Sketch triangle $ADC$.

$$\sin 70 = \frac{8}{AC}$$

$$AC = \frac{8}{\sin 70}$$

$$AC = 8.51...$$

Use trigonometry to find side $AC$.

Sketch triangle $ABC$.

> **WATCH OUT**
>
> The question does not give you enough information about triangle $ABC$ for you to use the sine or cosine rule straight away.

## 25 Trigonometry in 3D, sine and cosine rules

### Area of a triangle

area of a triangle $= \frac{1}{2} ab \sin C$ where $C$ is the angle between sides $a$ and $b$.

---

**Worked example**

Work out the area of this triangle.

8.1 cm

82°

12.9 cm

14.2 cm

area $= \frac{1}{2} \times 8.1 \times 12.9 \times \sin 82$

$= 51.7 \text{ cm}^2$ (1 d.p.)

> Find the two sides that the marked angle is between.
>
> Substitute these values into the formula.

---

**Worked example**

The area of triangle $ABC$ is 128 cm².

Work out the length of $AC$.

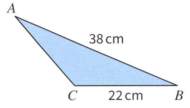

$A$

38 cm

$C$    22 cm    $B$

$128 = \frac{1}{2} \times 22 \times 38 \sin(\angle ABC)$

$\sin(\angle ABC) = \dfrac{8}{\frac{1}{2} \times 22 \times 38}$

$= 0.306\ldots$

$(\angle ABC) = \sin^{-1}(0.306\ldots)$

$= 17.8°$

> Substitute the known values into the formula for area of a triangle using the sine rule.

Using the cosine rule gives

$AC^2 = 22^2 + 38^2 - 2 \times 22 \times 38 \times \cos(17.8\ldots)$

$= 336.3$

$AC = \sqrt{336.3\ldots}$

$= 18.3 \text{ cm}$ (1 d.p.)

> Substitute the known values into the cosine rule equation.

---

**Key terms** Make sure you can write a definition for these key terms

cosine rule    Pythagoras' theorem    sine rule

## 25 Trigonometry in 3D, sine and cosine rules

Learn the answers to the questions below, then cover the answers column with a piece of paper and write as many as you can. Check and repeat.

| | Questions | Answers |
|---|---|---|
| 1 | State Pythagoras' theorem. | $a^2 = b^2 + c^2$ |
| 2 | What is the formula $a^2 + b^2 + c^2 = d^2$ used for? | To find the longest diagonal in a cuboid. |
| 3 | In the previous formula, what are $a$, $b$, and $c$? | The sides of the cuboid. |
| 4 | What is the sine rule to find an angle? | $\dfrac{\sin A}{a} = \dfrac{\sin B}{b} = \dfrac{\sin C}{c}$ |
| 5 | What is the sine rule to find a length? | $\dfrac{a}{\sin A} = \dfrac{b}{\sin B} = \dfrac{c}{\sin C}$ |
| 6 | What is the cosine rule to find a side? | $a^2 = b^2 + c^2 - 2bc \cos A$ |
| 7 | What is the cosine rule to find an angle? | $\cos A = \dfrac{b^2 + c^2 - a^2}{2bc}$ |
| 8 | How do you label a related side and an angle? | A related side and angle are opposite each other. |
| 9 | What is the formula to find the area of a triangle using trigonometry? | $\text{area} = \dfrac{1}{2} ab \sin C$ |
| 10 | The sine rule states that $\dfrac{\sin A}{a} = \dfrac{\sin B}{b}$. Rearrange the equation to make $A$ the subject. | $A = \sin^{-1}\left(a \sin \dfrac{B}{b}\right)$ |

*Put paper here*

### Previous questions

Now go back and use these questions to check your knowledge of previous topics.

| | Questions | Answers |
|---|---|---|
| 1 | How do we know when a ratio is written in its simplest form? | The only remaining common factor of each of the parts of the ratio is 1 |
| 2 | What is the general equation of a quadratic equation? | $y = ax^2 + bx + c$ |
| 3 | Give the exact value of tan 30° | $\dfrac{1}{\sqrt{3}}$ (which can also be written as $\dfrac{\sqrt{3}}{3}$ ) |
| 4 | What is the term-to-term rule? | The rule that gets you from one term to the next, which is the same for every term. |
| 5 | How do you solve a quadratic equation that can be factorised? | Rearrange the equation such that $ax^2 + bx + c = 0$, then factorise. |

*Put paper here*

# Practice

**25.1** Here is a cuboid.

The length of the longest diagonal of this cuboid is 13 cm.

The cuboid has length 12 cm and width 3 cm.

Work out the height, $h$, of the cuboid. **[3 marks]**

> **EXAM TIP**
>
> Use Pythagoras' theorem to write an equation for the diagonal.

**25.2** The 3D shape is made using 1 cm cubes. Vertices $P$ and $Q$ are labelled. Work out the length of the line $PQ$. **[2 marks]**

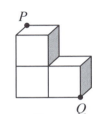

**25.3** $ABCDE$ is a square-based pyramid.

$E$ is vertically above the centre of the base of the pyramid.

The square base has a side length $\sqrt{2}$ cm and $AE$ is $\sqrt{3}$ cm. Calculate the height of the pyramid. **[3 marks]**

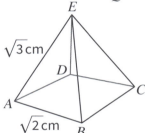

**25.4** $VWXY$ is a tetrahedron.

Its faces are equilateral triangles of side length 3 m.

$Y$ is directly above the centre of the base of the tetrahedron.

Work out the angle between the $XY$ and the base of the tetrahedron. Give your answer to the nearest degree. **[4 marks]**

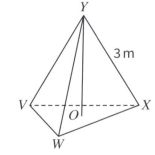

**25.5** *PQRS* is the horizontal rectangular base.

*T* is the point on *SR* such that *ST* : *TR* = 7 : 3

Work out the size of the angle between *OT* and the base of the prism.

Give your answer correct to 3 significant figures. **[5 marks]**

**25.6** The diagram shows a cuboid.

*O* is a point at the centre of the base of the cuboid

*FC* = 5 cm, *AB* = 4 cm and *HE* = 6 cm

Work out the angle between *FO* and the base of the cuboid.

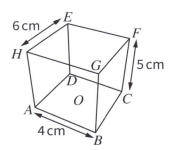

**[5 marks]**

**25.7** *ABC* is a triangle. **[6 marks]**

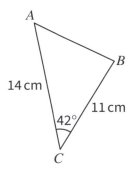

Work out the length of *AB*.

Give your answer correct to 1 decimal place.

**25.8** *DEF* is a triangle. **[6 marks]**

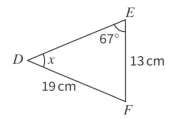

Work out the size of the angle marked *x*.

Give your answer correct to 1 decimal place. **[3 marks]**

# Exam-style questions

**25.9** A triangle has vertices $A$, $B$ and $C$.

$AB = 10$ cm, $BC = 6$ cm and angle $CAB$ is $25°$

Given that angle $ACB$ is obtuse, work out the size of angle $ACB$ to 1 decimal place. **[4 marks]**

**EXAM TIP**

Draw a sketch.

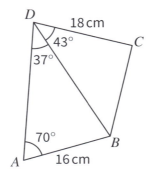

**25.10** Work out the length of $BC$.
Give your answer to 1 decimal place. **[5 marks]**

**25.11** A triangular lawn has sides of length 13 m, 11 m and 4 m.

Work out the area of the lawn correct to 3 significant figures. **[4 marks]**

**25.12** A triangular-based pyramid has of four equilateral triangular faces with side length 8 cm.

Work out the exact surface area of the pyramid.
State your units. **[4 marks]**

**25.13** Work out the area of quadrilateral $ABCD$. Give your answer correct to 1 decimal place. **[5 marks]**

**25.14** The diagram shows the intersection of a circle, centre $O$, and an equilateral triangle, $OAB$.

$AB = 8$ cm

$C$ is the midpoint of $OB$ and $D$ is the midpoint of $OA$.

**(a)** Work out the exact area of the shaded region $ABCD$. **[4 marks]**

**(b)** A chord is drawn on the circle from $C$ to $D$. Work out, to 1 decimal place, the area of the minor segment formed. **[3 marks]**

**25.15** *ABC* and *BCD* are triangles.
The area of triangle *ABC* is 80 cm².

Work out the size of angle *CBD*.
Give your answer to 3 significant
figures.

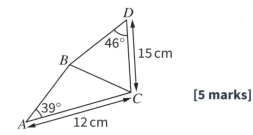

**[5 marks]**

**Questions referring to previous content**

**25.16** Rick writes a 2-digit odd number.

**(a)**   Work out how many different numbers Rick could write.   **[2 marks]**

**(b)**   What is the probability that he writes a number whose
digits are both the same? (Based on ALL two-digit numbers (10 to 99),
not just the odd ones found in part a.)   **[2 marks]**

 **25.17** Triangle *ABC* has vertices *A*(0, 0), *B*(2, −2) and *C*(−1, −4)

Work out the vertices of *A'B'C'* when *ABC* is reflected in the line $x = 1$   **[2 marks]**

## Circle theorems

Angles in the same **segment** are equal.

An angle in a **semicircle** is a right angle.

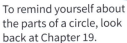

**LINK**

To remind yourself about the parts of a circle, look back at Chapter 19.

An angle at the centre is twice the angle at the **circumference**.

Opposite angles in a **cyclic quadrilateral** add up to 180°.

**REVISION TIP**

All the vertices of a cyclic quadrilateral lie on the circumference of a circle.

If you are asked to prove a **circle theorem**, use angle facts and other circle theorems to clearly justify your working at each stage.

### Worked example

$A$, $B$ and $C$ are points on a circle.

The centre of the circle is $O$.

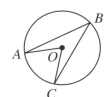

**(a)** Prove that $\angle AOC = 2 \angle ABC$

$AO = OB$, since both $AO$ and $OB$ are radii.

Hence, $\triangle ABO$ is isosceles.

Split $ABCO$ into $\triangle ABO$ and $\triangle CBO$.

State the angle facts and circle theorems that apply to $\triangle ABO$.

If $\angle OBA = x$, then $\angle OAB = x$, since base angles of an isosceles triangle are equal.

$\angle AOB = 180 - 2x$, since angles in $\triangle ABO$ sum to 180°.

Label $\angle OBA$ as $x$.

Consider $\triangle ABO$.

$\angle DOA = 180 - \angle AOB$

$\quad = 180 - (180 - 2x)$

$\quad = 2x$

Use triangle facts to work out $\angle DOA$.

$\angle DOC = 2y$ (as above)

$\angle ABC = x + y$

$\angle AOC = 2x + 2y$

$\quad = 2(x + y)$

$\quad = 2\angle ABC$

Use triangle facts to work out $\angle AOC$.

# Circle theorems

## Worked example

**(b)** $A$, $B$ and $C$ are points on a circle, centre $O$.

Angle $ABC = 35°$

Work out the size of $\angle CAO$.

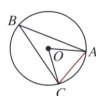

$\angle AOC = 2 \times \angle ABC$ — Angle at the centre is twice the angle at the **circumference**.

$= 2 \times 35°$

$= 70°$

In $\triangle CAO$, $AO = AC$ — $AO$ and $AC$ are radii and all radii in a circle must be equal.

Hence $\triangle CAO$ is isosceles.

$\angle CAO = \dfrac{180 - 70}{2}$ — The value of $\angle AOC$ is $70°$ and two angles of an isosceles triangle must be equal.

$= 55°$

The angle between a tangent and the radius is 90°.

Two tangents from the same point are equal in length.

The angle between a tangent and a **chord** is equal to the angle from the chord in the **alternate segment** of the circle.

## Worked example

Work out the sizes of angles $x$ and $y$.

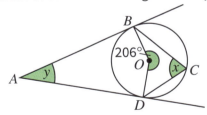

obtuse $\angle BOD = 360 - 206$

$= 154°$

$x = \dfrac{154}{2} = 77°$

angle at centre = angle at circumference $\times$ 2

$\angle ABO = \angle ADO = 90°$ (angle between tangent and radius is 90°)

$y = 360 - 90 - 90 - 154$

$= 26°$ (angles in a quadrilateral add up to 360°)

# ⚙ Knowledge

## 26 Circle theorems and circle geometry

## Circle geometry

The equation of a circle with centre (0, 0) and radius $r$ is

$$x^2 + y^2 = r^2$$

This comes from Pythagoras' theorem.

### Worked example

**1.** Write the equation of the circle shown.

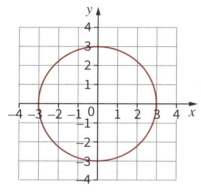

The centre of the circle is the origin and its radius is 3, so its equation is

$$x^2 + y^2 = 9$$

> Substitute 3 in place of $r$

**2.** Show that the point $(-1, 2\sqrt{2},)$ lies on the circle.

$$x^2 + y^2 = (-1)^2 + (2\sqrt{2})^2$$
$$= 1 + 8$$
$$= 9$$

so $(-1, 2\sqrt{2})$ lies on the circle.

> Substitute the coordinates into the equation.

> If the answer is 9, the point lies on the circle.

**LINK**

To remind yourself about multiplying surds, look back at Chapter 3.

## Equation of the tangent to a circle at a point on the circumference

The equation of the tangent to a circle at a point on the circumference is $y = mx + c$, where:
$m =$ the gradient of the line   $c =$ the intercept of the line at $y$-axis

### Worked example

A circle with centre (0, 0) has equation $x^2 + y^2 = 100$

The point $A$ (6, − 8) lies on the circle.
Work out the equation of the tangent to the circle at $A$.

gradient of $OA = -\dfrac{8}{6} = -\dfrac{4}{3}$

> Find the gradient of the radius $OA$.

gradient of tangent at

$$A = \frac{3}{4}$$

> Use $m_1 m_2 = -1$ for perpendicular lines to find gradient of tangent at $A$.

$$y = mx + c$$
$$-8 = \frac{3}{4} \times 6 + c$$

> Substitute $m = \dfrac{3}{4}$ and $(x, y) = (6, -8)$ into $y = mx + c$ to find $c$.

$$c = -\frac{25}{2}$$

$$y = \frac{3}{4}x - \frac{25}{2}$$

> Write the equation of the tangent.

## Equation of the tangent to a circle at a point on the circumference

**Worked example**

Solve these simultaneous equations.

Show your working.

$2x + y = 10$ ①

$x^2 + y^2 = 40$ ②

$y = 10 - 2x$ ③

$x^2 + (10 - 2x)^2 = 40$

$x^2 + 100 - 40x + 4x^2 = 40$

$5x^2 - 40x + 60 = 0$

$x^2 - 8x + 12 = 0$

$(x - 2)(x - 6) = 0$

$x = 2$ or $x = 6$

When $x = 2$,

$\quad y = 10 - 2(2)$

$\quad = 6$

When $x = 6$,

$\quad y = 10 - 2(6)$

$\quad = -2$

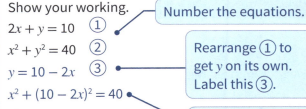

Number the equations.

Rearrange ① to get $y$ on its own. Label this ③.

Substitute $y$ into ②.

Solve for $x$.

Substitute the two values of $x$ into ③ to find $y$.

**REVISION TIP**

This solution tells you that the graphs of

$2x + y = 10$ and

$x^2 + y^2 = 40$ intersect

at the points $(2, 6)$ and $(6, -2)$.

**WATCH OUT**

Double check your solutions by substituting them into one of the other equations and checking that it balances.

---

**Worked example**

The graph of $x^2 + y^2 = 4$ is shown.

Use a graphical method to work out an estimate for the solutions to the simultaneous equations:

$x^2 + y^2 = 4$ ①

$y = 1 - \dfrac{1}{2}x$ ②

$x = 2, y = 0$ and $x = -1.2, y = 1.6$

Number the equations.

Draw the graph of $y = 1 - \dfrac{1}{2}x$ on the same axes.

Estimate the coordinates for the two points where the graphs intersect.

 **Make sure you can write a definition for these key terms**

alternate segment   chord   circle theorem
circumference   cyclic quadrilateral   segment   semicircle

## 26 Circle theorems and circle geometry

Learn the answers to the questions below, then cover the answers column with a piece of paper and write as many as you can. Check and repeat.

| Retrieval | | Answers |
|---|---|---|
| 1 | Write the relevant theorem/fact for this diagram. | Angles in the same segment are equal. |
| 2 | Opposite angles in a cyclic quadrilateral add up to what? | 180° |
| 3 | An angle at the centre is twice the angle of what? | The circumference. |
| 4 | What is the angle in a semicircle? | A right angle. |
| 5 | Two tangents from the same point are equal in what? | Length. |
| 6 | What is the angle between a tangent and the radius? | 90° |
| 7 | The angle between a tangent and a chord is equal to what? | The angle from the chord in the alternate segment of the circle. |
| 8 | What is the general equation of a circle that has its centre on the origin? | $x^2 + y^2 = r^2$ |

*Put paper here*

## Previous questions

Now go back and use these questions to check your knowledge of previous topics.

| Questions | | Answers |
|---|---|---|
| 1 | What does $2a$ mean? | 2 multiplied by $a$ |
| 2 | How can you check your solutions to a pair of simultaneous equations? | Substitute your values back into the equations. |
| 3 | What is the area of a 2D shape? | The space inside a 2D shape. |
| 4 | How do you find the volume of a prism? | Multiply the area of the cross-section by the length. |
| 5 | What units do you use for volume? | Cubic units. |

*Put paper here*

## Exam-style questions

**26.1** The circle $x^2 + y^2 = 1$ has centre at $(0, 0)$ and radius 1.
Lin says that the circle must pass through the point $(1, 1)$.

Show that Lin is wrong. **[2 marks]**

**26.2** The graph shows a circle, centre $(0, 0)$.

(a) Write the equation of the circle. **[1 mark]**

(b) Show that the point $(2\sqrt{2}, 2\sqrt{2})$
lies on the circle. **[2 marks]**

(c) Write the equation of the
tangent to the circle that
passes through the point $(0, -4)$. **[1 mark]**

**26.3** The circle, $C$, has equation $x^2 + y^2 = 5$

Work out the equation of the tangent to $C$ at the point $(1, 2)$. **[5 marks]**

**EXAM TIP**

Always draw a sketch if
you're not given one.

**26.4** A graph has equation $x^2 + y^2 = 10$

The graph passes through the points $A$, $B$, $C$ and $D$, which lie on the axes.

Prove that $ABCD$ is a square. **[4 marks]**

 **26.5** Write the size of angle $x$ in the diagram.

Give a reason for your answer. **[2 marks]**

 **26.6** (a) Write the size of angle $y$ in the
diagram, stating the circle
theorem used. **[2 marks]**

(b) Prove that angles in the same
segment of a circle are equal.
You may use, without proof,
the circle theorem you used
in part **a**. **[4 marks]**

**26.7** Work out the size of angle $z$. You must give a reason for each stage of your working. **[5 marks]**

 **26.8** Write the value of $x$ in the diagrams.

Give a reason for your answer. **[2 marks]**

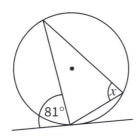

**26.9** $B$, $C$ and $D$ are points on a circle.

$AB$ and $AC$ are tangents.

$DB = DC$

Work out the size of angle $p$. **[4 marks]**

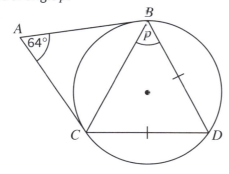

**26.10** A circle has equation $x^2 + y^2 = 25$

(a) Show that point $A$ with coordinates $(-4, 3)$ lies on this circle. **[1 mark]**

(b) Sketch the graph of the circle,

Label the coordinates where the circle intercepts the axes. **[2 marks]**

(c) A right-angled triangle, $ABC$, has each of its vertices on the circle. The coordinates of $A$ and $B$ are $(-4, 3)$ and $(3, 4)$ respectively.

Vertex $C$ has a negative $y$-coordinate

Work out two possible pairs of coordinates for $C$. **[3 marks]**

**26.11** Prove algebraically that the straight line with equation
$x + y + 6 = 0$ is a tangent to the circle $x^2 + y^2 = 18$ **[4 marks]**

**26.12** $OAC$ is a sector of a circle, centre $O$.

$AB$ is a tangent to the circle at point $A$.

$CB$ is a tangent to the circle at point $C$.

Calculate the area of the shaded region.
Give your answer to 3 significant figures. **[4 marks]**

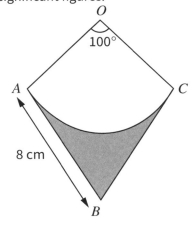

**26.13** $V$, $W$ and $X$ are points of the circumference of a circle, centre $O$.

$YXZ$ is the tangent to the circle at the point $X$.

Work out the size of angle $OVW$. **[4 marks]**

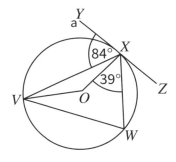

**Questions referring to previous content**

**26.14** The first three terms of a geometric sequence are $x + 2$, 6, $9x + 3$, where $x = 0$
Work out the exact value of the 5th term of the sequence. **[5 marks]**

**26.15** Triangle $ABC$ has sides of lengths
$AB = 8.8$, $BC = 7.8$ and $AC = 5.6$.

Work out the angle at vertex $A$ to the nearest tenth of a degree. **[3 marks]**

# ⚙ Knowledge

## 27 Vectors

## Vector notation

A **vector** is a way of describing movement.
Vectors have both **direction** and **magnitude** (size).

One way to write a vector is as a **column vector**. ————————●

Other notations for representing vectors include:

| **An arrow** | **A bold letter** | **An underlined letter** |
|---|---|---|
|  |  |  |

In the example above, the vector $\overrightarrow{BA}$ could be written as $-\mathbf{a}$ or $-\underline{a}$.
It is a parallel vector of the same length but opposite direction.

**LINK** 🔗

You revised translations of shapes using column vectors in Chapter 23. The top number tells you how far right or left the shape moves, and the bottom number tells you how far up or down it moves.

## Multiplying a vector by a scalar

A **scalar** is a quantity with just size. A number is a scalar.

Scalars can be used as scale factors or multipliers.

To multiply column vectors by a scalar, multiply the top and bottom values by the number.

For example:

$$3\begin{pmatrix} -1 \\ 8 \end{pmatrix} = \begin{pmatrix} 3 \times -1 \\ 3 \times 8 \end{pmatrix} = \begin{pmatrix} -3 \\ 24 \end{pmatrix}$$

**Worked example**

$$\mathbf{a} = \begin{pmatrix} 2 \\ 3 \end{pmatrix}$$

Draw vectors **a** and 2**a** on a grid.

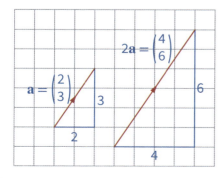

2 is a scalar. It only has magnitude (size).

$$2\mathbf{a} = 2\begin{pmatrix} 2 \\ 3 \end{pmatrix} = \begin{pmatrix} 2 \times 2 \\ 2 \times 3 \end{pmatrix} = \begin{pmatrix} 4 \\ 6 \end{pmatrix}$$

a is a vector. It has both direction and magnitude.

Notice how **a** and 2**a** are parallel.

## Parallel vectors

On this diagram, vector $\overrightarrow{CD}$ is also **a** as it has the same size and direction as the vector $\overrightarrow{AB}$.

Two vectors are **parallel** if one is a multiple of another.

The vector –**a** is the same size as **a** and parallel to **a** because it is $-1 \times$ **a**, but it is in the opposite direction to **a**.

A vector that is parallel to **a** but a different length will be a multiple of **a**, such as 2**a**.

### Worked example

$OABC$ is a parallelogram.

Not drawn accurately

**REVISION TIP** ✓
If two vectors are parallel, one is a multiple of the other.

**REVISION TIP** ✓
Remember to underline the letter if you write vectors by hand.

$\overrightarrow{OA} = $ **a** and $\overrightarrow{OC} = $ **c**

Write down these vectors in terms of **a** and **c**

$\overrightarrow{CO} \quad = -\underline{c}$

> it is parallel to $\overrightarrow{OC}$ but in the opposite direction.

$\overrightarrow{OB} \quad = \overrightarrow{OA} + \overrightarrow{AB} = \underline{a} + \underline{c}$

$\overrightarrow{AC} \quad = \overrightarrow{AO} + \overrightarrow{OC} = -\underline{a} + \underline{c}$

> you could write $\underline{c} - \underline{a}$, the order doesn't matter.

Points that lie on a straight line are collinear.

If $M$ is the midpoint of $AB$, then $\overrightarrow{AM} = \frac{1}{2}\overrightarrow{AB}$.

If two lines are parallel and both pass through a common point, they must lie on a straight line.

To show that points $A$, $B$ and $C$ lie on a straight line, you need to show that any two of $\overrightarrow{AB}$, $\overrightarrow{AC}$ or $\overrightarrow{BC}$ are multiples of each other.

### Worked example

$ABCD$ is a trapezium.
$\overrightarrow{AB} = 2\overrightarrow{DC}$ and $\overrightarrow{AF} = 2\overrightarrow{AB}$
$E$ is a point on $\overrightarrow{BC}$ such that $\overrightarrow{CE} : \overrightarrow{EB} = 1 : 2$
Show that $DEF$ is a straight line.

$\overrightarrow{DE} = \overrightarrow{DC} + \frac{1}{3}\overrightarrow{CB}$ since $CE : EB = 1 : 2$

$\quad = q + \frac{1}{3}(-q + p + 2q)$

> $\overrightarrow{AB} = 2\overrightarrow{DC} = 2q$

$\quad = \frac{1}{3}p + \frac{4}{3}q$

$\overrightarrow{DF} = \overrightarrow{DA} + 2\overrightarrow{AB}$

> $\overrightarrow{AF} = 2\overrightarrow{AB}$

$\quad = p + 2(2q)$

$\quad = p + 4q$

$\quad = 3\overrightarrow{DE}$

## 27 Vectors

## Adding and subtracting vectors

To add column vectors:
- add the top values together
- add the bottom values together.

**Worked example**

Given that $\mathbf{a} = \begin{pmatrix} 4 \\ -3 \end{pmatrix}$ and $\mathbf{b} = \begin{pmatrix} 1 \\ 7 \end{pmatrix}$

**1.** Work out $\mathbf{a} + \mathbf{b}$

$$\begin{pmatrix} 4 \\ -3 \end{pmatrix} + \begin{pmatrix} 1 \\ 7 \end{pmatrix}$$

$$\begin{pmatrix} 4+1 \\ -3+7 \end{pmatrix} = \begin{pmatrix} 5 \\ 4 \end{pmatrix}$$

Choose a starting point and draw vector **a**

**2.** Draw the vectors **a**, **b** and $\mathbf{a} + \mathbf{b}$ on a squared grid.

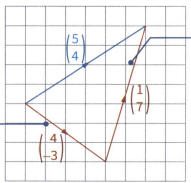

Starting from the end point of the first vector, draw vector **b**

The sum of the two vectors is the single vector that takes you from the starting point of the first vector to the end point of the second vector.

To subtract column vectors:
- subtract the top value of the second vector from the top value of the first
- subtract the bottom value of the second vector from the bottom value of the first.

## Using simultaneous equations

You can form a pair of simultaneous equations to solve for $x$ and $y$.

**Worked example**

$4\begin{pmatrix} x \\ 1 \end{pmatrix} = \begin{pmatrix} -y \\ 2x \end{pmatrix}$ Work out the values of $x$ and $y$.

$4x = -y$ ① 

$4 = 2x$ ②

Equate the top components and bottom components, and write as a pair of simultaneous equations.

$\dfrac{4}{2} = \dfrac{2x}{2} \Rightarrow x = 2$

From ②

$4 \times 2 = -y \Rightarrow y = -8$

Substitute $x = 2$ in ①

$x = 2, y = -8$

 **Key terms** | Make sure you can write a definition for these key terms

| column vector   direction   magnitude
| parallel   scale factor   vector

## 27 Vectors

Learn the answers to the questions below, then cover the answers column with a piece of paper and write as many as you can. Check and repeat.

| Questions | Answers |
|---|---|
| **1** What does magnitude mean? | Magnitude means size. |
| **2** What is a scalar? | A quantity that has only size. |
| **3** What are the different ways that this vector can be written?  | $\overrightarrow{AB}$ or **a** or <u>a</u> or $\begin{pmatrix} 5 \\ -3 \end{pmatrix}$ |
| **4** How do you know when vectors are parallel? | They are multiples of each other. |
| **5** What do you know about the relationship between **a** and −**a**? | −**a** is the same size as **a** and is parallel to **a**, but is in the opposite direction. |
| **6** What do you know about the relationship between **a** and 2**a**? | 2**a** is parallel to **a**, and in the same direction, but it is twice as long. |
| **7** What does the top value in a column vector tell you? | How far to move to the right, if the number is positive and how far to, move to the left if the number is negative. |
| **8** How do you add column vectors? | Add the top values together and add the bottom values together. |
| **9** How do you multiply a column vector by a scalar? | Multiply both the top and bottom values by the number. |
| **10** How can you represent the difference of two vectors, **a** − **b** on a grid? | −**b** is the same as **b**, but in the opposite direction. **a** − **b** is drawn as the sum of **a** and −**b**. |
| **11** How can you show that points $A$, $B$ and $C$ lie on a straight line? | Show that any two of $\overrightarrow{AB}$, $\overrightarrow{BC}$ and $\overrightarrow{AC}$ are parallel. |

*Put paper here*

## Previous questions

Now go back and use these questions to check your knowledge of previous topics.

| Questions | Answers |
|---|---|
| **1** What are the four types of transformations? | Reflections, rotations, translations, enlargements. |
| **2** What is a mixed number? | A number that has a whole number and a fraction. |
| **3** What is an inequality | A relationship between two expressions or values that are not equal to each other. |
| **4** What are the two algebraic methods you can use to solve simultaneous equations? | Elimination and substitution. |

*Put paper here*

# Practice

**27.1** The vector **a** and the vector **b** are shown on the grid.

    **(a)** On the grid, draw **a** − **b**.     **[2 marks]**

    **(b)** Work out 2**b** + 3**a** as a column vector.     **[2 marks]**

**27.2** *PQRSTU* is a regular hexagon. Write these vectors in terms of **a** and **b**.

    **(a)** $\overrightarrow{OT}$     **(b)** $\overrightarrow{PQ}$

    **(c)** $\overrightarrow{OU}$     **(d)** $\overrightarrow{UQ}$     **[4 marks]**

**27.3** **p** and **q** are vectors.

$\mathbf{p} = \begin{pmatrix} 4 \\ -1 \end{pmatrix}$ and $2\mathbf{q} - 5\mathbf{p} = \begin{pmatrix} -26 \\ 15 \end{pmatrix}$

    Write **q** as a column vector.     **[3 marks]**

**27.4** In the grid shown, $\overrightarrow{OA} = \begin{pmatrix} 0 \\ 2 \end{pmatrix} \overrightarrow{OB} = \begin{pmatrix} 2 \\ 4 \end{pmatrix}$ and $\overrightarrow{OD} = \begin{pmatrix} 1 \\ 1 \end{pmatrix}$

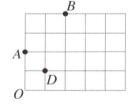

    *ABCD* is a rectangle.

    Write the column vector that represents $\overrightarrow{OC}$.     **[2 marks]**

**27.5** $\overrightarrow{OA} = \mathbf{a}$ and $\overrightarrow{OB} = \mathbf{b}$

    *M* is the midpoint of $\overrightarrow{OA}$

    $AN : NB = 4 : 1$

    Work out $\overrightarrow{MN}$ in terms of **a** and **b**.     **[4 marks]**

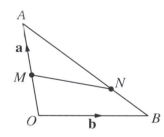

**27.6** $\overrightarrow{OP} = \mathbf{p}$ and $\overrightarrow{OQ} = \mathbf{q}$

    *PQR* is a straight line.

    $PQ : QR = 2 : 3$

    Show that $\overrightarrow{OR}$ is parallel to 5**q** − 3**p**     **[5 marks]**

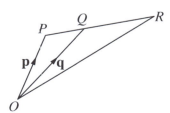

> **EXAM TIP**
>
> Parallel vectors are a constant multiple of each other.

**27.7** *OAB* is a triangle.

*OPA* and *OQB* are straight lines.

$\overrightarrow{OA} = \mathbf{a}$ and $\overrightarrow{OB} = \mathbf{b}$.

*M* is the midpoint of *QA*.

$OP : PA = 3 : 1$

Work out the ratio $OQ : QB$.    **[5 marks]**

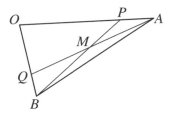

**27.8** *OABC* is a trapezium with *OA* parallel to *CB*.

*OPC* and *APD* are straight lines.

$\overrightarrow{OA} = 9\mathbf{a}$ and $\overrightarrow{OC} = \mathbf{c}$.

$AP : PC = 3 : 1$.

Prove that *O*, *P* and *B* are collinear.    **[4 marks]**

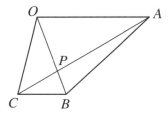

**Questions referring to previous content**

**27.9** At time $t = 0$, Riley passes point *A*, running at a constant speed of 8 km/h. At the same instant, Jordan, running in the opposite direction, passes point *B* at a constant speed of 2.5 m/s.

The distance between *A* and *B* is 8500 m.

8 km/h

Riley

2.5 m/s

Jordan

*A* ←———— 8500 m ————→ *B*

**(a)** Which of the runners is travelling faster?

Show working to support your answer.    **[2 marks]**

**(b)** Work out the time in minutes when the runners pass each other.    **[3 marks]**

**(c)** Work out how far the runners will be from *A* when they pass each other.    **[2 marks]**

**27.10** Prove that the angle in a semicircle is a right angle.

Do not use circle theorems in your proof.    **[4 marks]**

# Knowledge

## 28 Probability

## Probability basics

**Probability** is the likelihood that an event will happen.

It is always a value between 0 and 1.

Probabilities can be written as fractions, decimals or percentages.

**The probability scale**

## Theoretical probability

A **theoretical probability**, written as P(event), is the probability that an event will happen.

If all possible outcomes are equally likely, then the theoretical probability is:

$$P(\text{event}) = \frac{\text{number of ways event can happen}}{\text{total number of possible outcomes}}$$

For example, the probability of rolling a six with a dice is

$$P(\text{six}) = \frac{1}{6}$$

- There is one six on a dice.
- There are six numbers on a dice.

**Mutually exclusive** events are events that can't happen at the same time.

For example, a number cannot be both odd and even.

**Odd numbers**

3
1       5
7   9

**Even numbers**

2
10      4
8    6

The probabilities of all possible mutually exclusive outcomes must add up to 1.

P(event happening) = 1 − P(event not happening)

---

**Worked example**

In a set of shapes there are only triangles, squares, pentagons and hexagons.

A shape is chosen at random.

The table shows the probabilities of choosing a triangle or a square.

The ratio of pentagons to hexagons is 5 : 7

Complete the table.

| Shape | Triangle | Square | Pentagon | Hexagon |
|---|---|---|---|---|
| Probability | 0.38 | 0.26 | 0.15 | 0.21 |

$1 - (0.38 + 0.26) = 0.36$ — The four probabilities must add to 1.

$P(\text{pentagon}) : P(\text{hexagon}) = 5 : 7$

$P(\text{pentagon}) = \frac{5}{12} \times 0.36$ — Use the ratio of pentagons to hexagons to write the fraction for the number of pentagons. Multiply by the remaining probability.

$\quad = 0.15$

$P(\text{hexagon}) = \frac{7}{12} \times 0.36$ — Use the ratio of pentagons to hexagons to write the fraction for the number of hexagons. Multiply by the remaining probability.

$\quad = 0.21$

**LINK**

Look back at Chapter 16 to remind yourself about ratios.

# Outcomes and possibility spaces

The **possibility space** or **sample space** is a list or table showing all possible outcomes of an experiment or of combined experiments.

To show a possibility space:

- for one experiment, make a systematic list of all possible outcomes
- for two combined experiments, make a two-way table showing the combinations or end results.

### Worked example

As part of a board game, players must select a card from a pack and roll an ordinary dice.

Half of the cards have a number 2 on them and the rest have a number 3.

The player multiplies the number on the card by the number on the dice to give their score.

**1.** Draw a table to show the possibility space.

**Dice**

| card | 1 | 2 | 3 | 4 | 5 | 6 |
|------|---|---|---|----|----|----|
| **2** | 2 | 4 | 6 | 8 | 10 | 12 |
| **3** | 3 | 6 | 9 | 12 | 15 | 18 |

> Create a two-way table with the dice outcomes across the top, and card outcomes down the side.
>
> Multiply the numbers together to complete the possibility space.

**2.** Work out the probability of a player scoring more than 10.

There are 12 equally likely outcomes.

There are four outcomes which are more than 10.

$$P(\text{more then } 10) = \frac{4}{12}$$
$$= \frac{1}{3}$$

# Relative frequency

If you don't know the probability of an event, you can do an experiment and use your results to **estimate** the probability. This way of estimating probability from an experiment is called finding the **relative frequency**.

$$\text{relative frequency} = \frac{\text{number of times the event happened}}{\text{total number of trials}}$$

### Worked example

A biased dice is rolled 60 times.

> An unfair coin or spinner or dice is said to be **biased**.

20 of the rolls land on 5.

Estimate the probability of rolling 5 on Jamie's dice.

$$\text{estimate for } P(5) = \frac{\text{number of 5s}}{\text{total number of rolls}}$$

> Identify the event, and the total number of trials.

$$= \frac{20}{60} = \frac{1}{3}$$

> Substitute into the equation, and simplify the fraction.

 **REVISION TIP**

Increasing the number of times an experiment is repeated results in a better estimate for the real probability. This is because, with more repeats, the relative frequency gets closer and closer to the real probability.

## 28 Probability

## Expected results

The **expected frequency** of an event is the number of times you expect it to happen.

expected frequency = probability × number of trials

### Worked example

1. Bishal rolls a dice 200 times and gets a 6 on 30 rolls.
   Miles says' this means the dice is biased.
   Use data to give a reason why he might be incorrect.

   With a fair dice, $P(6) = \frac{1}{6}$ •———— Identify the expected probability.

   expected frequency $= \frac{1}{6} \times 200 = 33.333\ ...$ •—— Calculate the expected frequency.

   This is close to the actual result of 30, so Miles might be incorrect. The dice might be fair. •—— Compare the two and give a reason for your answer.

2. Ben rolls a dice six times and doesn't get a 6. He thinks that the dice is biased. Explain whether Ben is correct.
   Ben is wrong, because six rolls are not enough trials to draw a conclusion.

> **REVISION TIP**
> The expected frequency will not always match the actual outcome as it is just the most likely of many possible outcomes. If it is far off, though, the dice (or coin, spinner, etc.) is probably biased.

## Tree diagrams

**Tree diagrams** can be used when considering probability involving two or more events.

- To find the probability of two events occurring together, **multiply** probabilities along the branches.
- To find the probability of one **or** the other happening, **add** the probabilities along the branches.

### Worked example

There are only black dice and red dice in a bag.

Jo takes a dice at random from the bag and rolls it.

The **tree diagram** shows the probabilities.

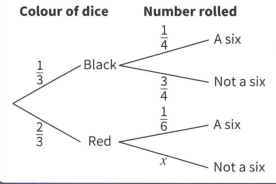

**Colour of dice**  **Number rolled**

1. What is the value of $x$?

   $x = 1 - \frac{1}{6}$ •———— Probabilities on each pair of branches must add up to 1.

   $= \frac{5}{6}$

2. Work out the probability of picking a black dice and rolling a 6.

   $P(\text{black and 6}) = \frac{1}{3} \times \frac{1}{4} = \frac{1}{12}$ •—— Multiply along the branches.

3. Work out the probability of rolling a 6.

   $P(\text{black and 6}) = \frac{1}{12}$ •—— Find the two possible ways to roll a 6.

   $P(\text{red and 6}) = \frac{2}{3} \times \frac{1}{6} = \frac{2}{18}$

   $P(\text{six}) = \frac{1}{12} + \frac{2}{18} = \frac{7}{36}$ •—— The two ways are mutually exclusive, so add them to find the probability of rolling a 6.

## Conditional probability

**Conditional probability** is the probability of one event happening, given that another event has already occurred.

$$P(A \mid B) = \frac{P(A \cap B)}{P(B)}$$

The denominator is always the probability of the event you *know* has happened. In this case, $B$ has occurred so the sample space is $B$.
The numerator is the probability of the event you are interested in: $A$.
The probability of this event happening in the sample space $B$ is $P(A \cap B)$: the probability of both $A$ and $B$ happening.

**REVISION TIP**

When a question says 'without replacement', it will always involve conditional probability. A question about conditional probability often uses the word 'given'.

> **Worked example**
>
> 60 people were asked how they travel to work.
>
> - 33 of the people were men.
> - 9 of the 23 people who travel by bus were women.
> - 12 men walk to work.
> - 19 people travel to work by car.
>
> **REVISION TIP**
>
> It is useful to write the information from a probability question in a two-way table.
>
> |       | bus | walk | car | total |
> |-------|-----|------|-----|-------|
> | men   | 14  | 12   | 7   | 33    |
> | women | 9   | 6    | 12  | 27    |
> | total | 23  | 18   | 19  | 60    |
>
> **1.** Calculate the probability that they travel by car and are men
>
> $$P(\text{car and men}) = \frac{7}{60}$$
>
> Here the sample space is all 60 people.
>
> **2.** Calculate the probability that they travel by car given they are men
>
> $$P(\text{car and given men}) = \frac{7}{33}$$
>
> Here the sample space is only the men.

## Venn diagrams

$\mathscr{E}$ means the **universal set**, which contains all the numbers.

$A \cap B$ means the **intersection** of sets $A$ and $B$ (numbers in **both** sets).

$A \cup B$ means the **union** of sets $A$ and $B$ (numbers in **either** set).

$A'$ means the **complement** of set $A$ (numbers **not** in set $A$).

$A \cap B$

$A \cup B$

$A'$

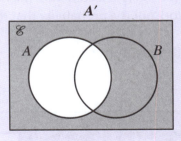

You can describe a set using curly brackets, for example, {1, 2, 3, 4} or {odd numbers}.

$x \in A$ means $x$ is a member of the set $A$.

# ⚙ Knowledge

## 28 Probability

### Venn diagrams

**Worked example**

There are 100 people at a swimming gala.

50 people are spectators.

There are three races: freestyle, backstroke and breaststroke.

10 people compete in all three races.

25 people compete in backstroke and breaststroke.

20 people compete in backstroke and freestyle.

12 people compete in freestyle and breaststroke.

30 people compete in freestyle.

1 person competes in only backstroke.

1. One of the people at the swimming gala is chosen at random.

   Work out the probability that the person competes in the breaststroke race.

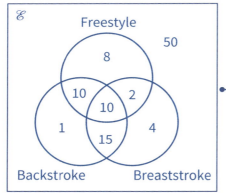

First draw a Venn diagram:

- identify the number of sets and draw three intersecting circles
- start in the middle with the numbers in all three sets
- complete the numbers in two sets
- complete the numbers in only one set
- complete the numbers in none of the sets.

$$P(\text{breaststroke}) = \frac{10 + 15 + 2 + 4}{100}$$

Add the numbers in the four segments of the breaststroke set and divide by the total number of people at the event.

$$= \frac{31}{100}$$

2. A person is selected from the backstroke racers.
   Work out the probability that they are also competing in freestyle.

$$P(\text{freestyle given backstroke}) = \frac{10 + 10}{10 + 10 + 15 + 1}$$

Restrict the sample space to people competing in backstroke.

$$= \frac{20}{36} = \frac{5}{9}$$

**Key terms** Make sure you can write a definition for these key terms

biased   conditional probability   complement   estimate
expected frequency   intersection   mutually exclusive
possibility space (or sample space)   probability
probability scale   relative frequency   theoretical probability
tree diagram   union   universal set

 **28** Probability

## 28 Probability

Learn the answers to the questions below, then cover the answers column with a piece of paper and write as many as you can. Check and repeat.

| Questions | Answers |
|---|---|
| 1 What do we call a list or table showing all possible outcomes of an experiment or of combined experiments? | A possibility space or a sample space. |
| 2 What is the estimated probability of an event happening called? | The relative frequency. |
| 3 How do you calculate the relative frequency of an event? | Number of times the event happened divided by total number of trials. |
| 4 How can you improve the accuracy of an estimated probability? | Increase the number of trials. |
| 5 What is the expected frequency of an event? | The number of times you expect a event to happen. |
| 6 How do you calculate probabilities along a branch of a tree diagram? | Multiply together the probabilities on the branch. |
| 7 What do the probabilities on a pair of branches of a probability tree add up to? | 1 |
| 8 If one event effects the probability of another happening what do we call this? | Conditional probability. |
| 9 What does $A \cup B$ mean? | $A$ or $B$ |
| 10 What does $A \cap B$ mean? | $A$ and $B$, the intersection of the Venn diagram. |
| 11 What does $A'$ mean? | The complement of $A$, not $A$. |

*Put paper here*

## Previous questions

Now go back and use these questions to check your knowledge of previous topics.

| Questions | Answers |
|---|---|
| 1 What does HCF stand for? | Highest common factor. |
| 2 What is ratio used for? | To compare two or more numbers. |
| 3 When you square a negative number what type of number will your answer be? | Positive. |
| 4 If $\tan \theta = 5$, how do you find $\theta$? | $\theta = \tan^{-1} 5$ |
| 5 In the number $5^3$, which value is the base? | 5 |

*Put paper here*

# Practice

## Exam-style questions

**28.1** A bag contains white, yellow, pink and orange counters. The table shows the probabilities of selecting each colour of counter from the bag.

| Colour | White | Yellow | Pink | Orange |
|---|---|---|---|---|
| Probability | 0.3 | 0.15 | 0.26 | |

    **(a)** Work out the probability of not selecting a white counter or an orange counter. **[2 marks]**

    **(b)** There are 20 counters in the bag. Work out how many orange counters there are. **[2 marks]**

**28.2** A bag contains 4 red cubes, 6 yellow cubes and 5 blue cubes.

Grace adds more red cubes to the bag until the probability of choosing a red at random is $\frac{1}{2}$.

How many red cubes does Grace add? **[2 marks]**

 **28.3** An unbiased 8-sided dice has the numbers 1, 2, 2, 3, 3, 4, 4 and 4 on its faces.

    **(a)** If the dice is thrown once, what is the probability that it will land on 4? **[1 mark]**

    **(b)** If the dice is thrown 40 times, how many times would you expect it to land on 3? **[2 marks]**

    **(c)** The dice is thrown repeatedly and lands on the number 4 a total of 36 times. Estimate how many times the dice was thrown. **[2 marks]**

**28.4** A manufacturer of light bulbs claims that 92% of light bulbs they produce last for longer than 25 000 hours.

The light bulbs are sold in packs of two.

A shop buys 500 packs.

Work out an estimate for the number of packs of light bulbs that will have exactly one light bulb that lasts longer than 25 000 hours. **[3 marks]**

> **EXAM TIP**
> It may help you to draw a tree diagram.

**28.5** The frequency tree shows the outcomes for 40 people who took a driving test.

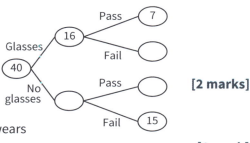

    **(a)** Complete the frequency tree. **[2 marks]**

    **(b)** Write the probability that a person chosen at random wears glasses and passes the test. **[1 mark]**

**28.6** When dropped onto a hard surface, a plastic cup will land the right way up, on its side or upside down.

**(a)** Sheena drops a plastic cup 50 times and records the results.
Complete the table. **[2 marks]**

|  | Right way up | On its side | Upside down |
|---|---|---|---|
| **Frequency** | 8 |  |  |
| **Relative frequency** |  | 0.44 |  |

**(b)** She then drops the plastic cup a further 100 times and it lands upside down 36 times.

Based on this, was the probability of the cup landing upside down higher in the first or second experiment? Show your working. **[2 marks]**

**28.7** Nasim plays a game of darts with her friend.

They then play a game of backgammon.

The tree diagram shows Nasim's probabilities of winning and losing each game.

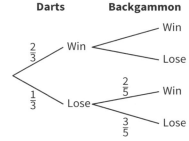

The probability that Nasim wins both games is $\frac{1}{2}$.

**(a)** Complete the two missing probabilities on the tree diagram. **[3 marks]**

**(b)** Work out the probability that Nasim's friend wins at most one of the two games. **[3 marks]**

**28.8** Two people are chosen at random from a group of 4 French speakers and 7 English speakers.

Using a tree diagram, show that the probability of selecting one French speaker and one English speaker is just over 50%. **[4 marks]**

**28.9** At a school there are a total of 150 pupils in either year 9 or year 10.

Each pupil studies one foreign language: French, German or Italian. Of the 87 pupils who study French, 35 are in year 9. 17 pupils in year 10 study German.

A total of 12 pupils study Italian.

There are 76 pupils in year 9.

One of the pupils is chosen at random.

Work out the probability that the pupil is in year 9, given that they study Italian. **[4 marks]**

# Exam-style questions

**28.10** A coin and a dice are repeatedly thrown together. Some of the results are shown in the two-way table.

EXAM TIP

Think about what the numerators and the denominators in the fractions represent.

| Dice<br>Coin | Even number | Odd number |
|---|---|---|
| **Heads** | | 23 |
| **Tails** | 24 | |

The probabilty of getting tails, given that an odd number is rolled on the dice, is $\frac{26}{49}$. The probability of rolling an odd number on the dice, given that you get heads on the coin, is $\frac{23}{50}$. Complete the missing two values in the table. **[2 marks]**

**28.11** The tree diagram shows the probabilities of a hockey player scoring a penalty on two consecutive attempts.

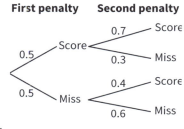

First penalty    Second penalty

(a) Work out the probability that the hockey player scores with the 2nd penalty. **[2 marks]**

(b) Work out the probability that the hockey player scores with the 1st penalty given that they miss the 2nd penalty. **[3 marks]**

EXAM TIP

$$P(A \text{ given } B) = \frac{P(A \text{ and } B)}{P(B)}$$

**28.12** $\mathscr{E} = \{1, 2, 4, 8, 16, 25, 27, 64\}$,

$A = \{\text{square numbers}\}$,

$B = \{\text{cube numbers}\}$.

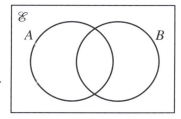

(a) Complete the Venn diagram. **[4 marks]**

(b) Work out $P(A \cap B)$. **[2 marks]**

**28.13** $\mathscr{E} = \{1, 2, 3, 5, 6, 9, 10, 11, 17, 21, 25\}$

$F = \{\text{prime numbers}\}$, $G = \{\text{numbers greater than 10}\}$ **[4 marks]**

(a) Draw a Venn diagram for this information. **[4 marks]**

(b) One of the numbers in the diagram is chosen at random. Work out

(i) $P(F \cap G)$ (ii) $P(G')$ (iii) $P(F)$ **[3 marks]**

**28.14** The Venn diagram shows the number of customers buying apples, pears and bananas.

EXAM TIP

'Given' means that the possibility space is restricted.

(a) A customer is chosen at random. Work out the probability that this customer does not buy apples.

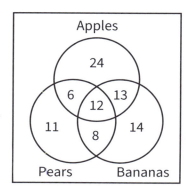

**[2 marks]**

**(b)**   A second customer is chosen at random.

Given that a randomly chosen customer bought bananas, work out the probability that they also bought apples?   **[2 marks]**

**28.15** Out of 200, people 48 people visited Africa. 5 visited each of Europe, North America and Africa. Of the 58 people who visited North America, 34 had not visited Europe or Africa.

16 people visited both Europe and North America, whilst 19 people had visited both Europe and Africa.

22 people visited none of the three continents.

**(a)**   A person is chosen at random. Work out the probability that a this person visited Europe.   **[4 marks]**

**(b)**   A second person is chosen at random. Given that a randomly chosen person visited North America, work out the probability that they also visited Africa.   **[2 marks]**

**28.16** There are only green discs, yellow discs and pink discs in a bag.

number of yellow discs : number of pink discs is $7:11$

Will takes a disc at random from the bag.

The probability that the disc is green is 0.4

Work out the probability that Will takes a pink disc.   **[3 marks]**

**28.17** There are only $b$ black shirts and $w$ white shirts on a rail.

A shirt is taken from the rail at random.

The probability that the shirt is black is $\frac{4}{9}$.

The shirt is then placed back on the rail.

4 more black shirts and 8 more white shirts are added to the rail.

A shirt is taken from the rail at random.

The probability that the shirt is black is $\frac{5}{12}$.

Work out the number of black shirts and the number of white shirts that were on the rail originally.   **[5 marks]**

**Questions referring to previous content**

**28.18 (a)**   $\mathbf{a} = \begin{pmatrix} 3 \\ -5 \end{pmatrix}$ and vector $\mathbf{b} = \begin{pmatrix} 9 \\ 4 \end{pmatrix}$

Work out $3\mathbf{a} + 2\mathbf{b}$   **[2 marks]**

**(b)**   Work out vector $\mathbf{c}$ if $\mathbf{b} - 3\mathbf{c} = 2$   **[3 marks]**

Simplify fully $2\sqrt{80} + 3\sqrt{50} + 4\sqrt{45}$   **[3 marks]**

## 29 Tables, averages, and range

## Averages

| Average | Description | Example |
|---|---|---|
| Mean | $\text{mean} = \dfrac{\text{sum of all values}}{\text{number of values}}$<br><br>The mean is affected by outliers (unusually large or small values). | The mean of 3, 4, 8, 5 is:<br>$\dfrac{3+4+8+5}{4} = \dfrac{20}{4}$<br>$= 5$ |
| Median | The middle value when the data is put in order.<br><br>If there are is an even number of values, there will be two 'middle values'. The median is the mean of these two values.<br><br>*Use the median instead of the mean if data has an outlier.* | The median of 3, 4, 5, 7 is:<br>$\dfrac{4+5}{2} = \dfrac{9}{2}$<br>$= 4.5$ |
| Mode | The most commonly occurring value.<br>There can be no mode, one mode or multiple modes.<br><br>*Use the mode if you have non-numerical data.* | The mode of<br>3, **4**, 7, 5, 3, **4**, 8, **4**, 6 is: 4 |
| Range | The difference between the largest and smallest values. | The range of 3, 4, 7, 5 is:<br>$7 - 3 = 4$ |

## Ungrouped frequency tables

You can find the mean, median, mode and range when data is in an ungrouped frequency table.

### Worked example

The table shows the number of sunny days per week over 21 weeks.

| Number of sunny days per week | 0 | 1 | 2 | 3 | 4 | 5 | 6 | 7 | Total |
|---|---|---|---|---|---|---|---|---|---|
| Frequency | 3 | 8 | 4 | 2 | 1 | 1 | 0 | 2 | 21 |
| No. days × frequency | 0 | 8 | 8 | 6 | 4 | 5 | 0 | 14 | 45 |

1. Calculate the median.

   0, 0, 0, 1, 1, 1, 1, 1, 1, 1, ①, 2, 2, 2, 2, 3, 3, 4, 5, 7, 7 ●——— Write out the data in an ordered list, and find the middle value(s).

2. Calculate the mode.

   The mode is 1. ●——— Find the number with the highest frequency in the table.

3. Calculate the mean.

   $\text{Mean} = \dfrac{45}{21}$ ●——— Add a row to the frequency table for number of days × frequency.

   $= 2.1\ (1\ \text{d. p.})$

4. Calculate the range.

   $\text{Range} = 7 - 0$ ●——— Then divide this total by the number of days to find the mean number of sunny days per week.

   $= 7$

   highest value – lowest value

> **REVISION TIP**
> To find the mean, always divide by the sum of all the frequencies.

# Grouped frequency tables

When data is given in a grouped frequency table, you can find:
- an estimate for the mean
- the **modal class** – the group with the highest frequency
- the **median class** – the group that the middle value lies in

For an ordered set of $n$ data values, the median is the $\frac{1}{2}(n+1)$th data value.

> **Worked example**
>
> The lengths of 25 pieces of string are recorded in the table.
>
> | Length of string $s$ (cm) | Frequency | Running total | Midpoint | Midpoint × frequency |
> |---|---|---|---|---|
> | $0 < s \le 8$ | 12 | 12 | 4 | 48 |
> | $8 < s \le 12$ | 6 | 18 | 10 | 60 |
> | $12 < s \le 20$ | 7 | 25 | 16 | 112 |
> | **Totals** | | | – | **220** |
>
> **1.** Which is the modal class?
>
> $0 < s \le 8$ is the modal class ⟵ Find the class with the highest frequency.
>
> ⟶ Add columns for the midpoint and midpoint × frequency, and calculate the values.
>
> **2.** Which class does the median lie in? ⟵ Add a 'running total' column to the table.
>
> There are 25 pieces of string, so the median will be the 13th. ⟵ Identify the position of the median.
>
> 13 lies between 12 and 18, so the 13th value lies in the $8 < s \le 12$ class. ⟵ See which class the 13th value lies in.
>
> **3.** Calculate an estimate for the mean.
>
> Estimate for mean $= \dfrac{220}{25}$ ⟵ Substitute the values into the equation for the mean.
>
> $= 88$
>
> ---
>
> **REVISION TIP**
>
> It is not possible to find the exact mean because we do not know all the exact values. So, estimate by assuming that, for each group, the length of each string is equal to the midpoint of the group.

**29** Tables, averages, and range

## Median, quartiles, and interquartile range

For data in ascending order:

**Median** – the value 50% of the way through the data
$$\frac{1}{2}(n+1)$$

0, 1, 2, 2, 3, 4, 4, 4, 4, 5, 5, 6, 7, 9, 9, 9

**Lower quartile** (LQ) – the value 25% of the way through the data
$$\frac{1}{4}(n+1)$$

**Upper quartile** (UQ) – the value 75% of the way through the data
$$\frac{3}{4}(n+1)$$

**Interquartile range** = upper quartile – lower quartile

---

### Worked example

Daisy played 11 games of football.
Here are the number of goals she scored in each game:

1  0  1  2  0  0  3  1  1  0  2

Find the interquartile range.

0, 0, 0, 0, 1, 1, 1, 1, 2, 2, 3 •———— Arrange the 11 items of data in ascending order.

$\frac{1}{4}(n+1) = 3 \Rightarrow$ 3rd value is 0 •———— Identify the position of the lower quartile.

$\frac{3}{4}(n+1) = 9 \Rightarrow$ 9th value is 2 •———— Identify the position of the upper quartile.

Interquartile range = 2 − 0

= 2 goals •———— Calculate the interquartile range.

Jessie also played 11 games of football.
Her median number of goals scores was 2 and her interquartile range was 3.
Compare Daisy's and Jessie's results.

$\frac{1}{2}(n+1) = 6 \Rightarrow$ 6th value is 1 •———— Calculate the position of the median.

Therefore, Daisy's median number of goals scored is 1.

Jessie's median is higher than Daisy's, •———— Compare the medians.
so Jessie scores more goals on average.

Daisy's interquartile range is smaller than Jessie's, •—— Compare the interquartile ranges.
so Daisy scores more consistently.

---

**Key terms** — Make sure you can write a definition for these key terms.

interquartile range    lower quartile    mean    median
median class    modal class    mode    range    upper quartile

## 29 Tables, averages, and range

Learn the answers to the questions below, then cover the answers column with a piece of paper and write as many as you can. Check and repeat.

| | Questions | | Answers |
|---|---|---|---|
| 1 | How would you find the mode from a frequency table? | Put paper here | Look for the interval or value with the highest frequency. |
| 2 | Do the groups need to be the same width in a grouped frequency table? | | No. |
| 3 | What is the median of a set of data? | | The middle value when the data is listed in order. |
| 4 | How do you calculate the range of a set of data? | Put paper here | Add up all the values and divide by the number of values. |
| 5 | How do you calculate the mean from a list of values? | | Add up all the values and divide by the number of values. |
| 6 | What is the modal class of a grouped frequency table? | Put paper here | The class that contains the mode. |
| 7 | What is the median class of a grouped frequency table? | | The class that the middle value lies in. |
| 8 | What should you do if there are two middle values when finding the median? | | Find the mean of the two middle values. |
| 9 | How do you calculate the mean of ungrouped data in a frequency table? | Put paper here | Multiply each value by its frequency, add up these values, and divide by the total of the frequencies. |
| 10 | What is an outlier? | | A value which doesn't fit the trend. |

## Previous questions

Now go back and use these questions to check your knowledge of previous topics.

| | Questions | | Answers |
|---|---|---|---|
| 1 | How do you represent an inequality on a number line where the value is NOT included? | Put paper here | Draw an open circle. |
| 2 | How do you find the gradient of a curve at a point? | | Draw a tangent and find its gradient. |
| 3 | What is the hypotenuse? | | The longest side of a right-angled triangle . |
| 4 | What is the general equation of a circle that has its centre at the origin? | | $x^2 + y^2 = r^2$ |
| 5 | What is the opposite of factorising? | | Expanding. |

# Practice

**29.1** The mean of ten numbers is 63. The mean of four of the numbers is 51.

Work out the mean of the remaining six numbers. **[3 marks]**

**29.2** 71 people take part in a 5-km race.
The table shows their finishing times in minutes.

| Time $t$ (minutes) | Frequency |
|---|---|
| $20 < t \le 25$ | 10 |
| $25 < t \le 30$ | 17 |
| $30 < t \le 35$ | 24 |
| $35 < t \le 40$ | 11 |
| $40 < t \le 45$ | 9 |

(a) Write the class interval that contains the median. **[2 marks]**

(b) Work out an estimate for the mean finishing time.
Give your answer to the nearest minute. **[3 marks]**

(c) Explain why your answer to part **b** is only an estimate. **[1 mark]**

**29.3**

| Score ($x$) | Frequency |
|---|---|
| $0 < x \le 4$ | $3y$ |
| $4 < x \le 8$ | $7y$ |

> **EXAM TIP**
> Insert extra columns in to the table.

Work out an estimate of the mean score. **[3 marks]**

**29.4** The interquartile range of the numbers on these cards is 3.

Work out the missing number? **[2 marks]**

 **29.5** Two basketball teams, A and B, each play 12 games.
Their scores are shown.

Team A:  87 85 82 76 76 100 95 63 123 78 90 99

Team B:  105 97 89 103 78 89 103 70 97 98 101 88

| | |
|---|---|
| Lowest Score | 63.0 |
| Lower Quartile | |
| Median | |
| Upper Quartile | |
| Highest Score | 123.0 |

Here is the information for Team B.

(a) Complete the table for Team A. **[2 marks]**

(b) Josh says, 'Team A have the highest score, so they must be the better team.'
Is Josh correct? Explain your answer. **[2 marks]**

(c) Taisa says, 'Team A's scores are more varied than Team B's scores.'
Is Taisa correct? Explain your answer. **[2 marks]**

| | |
|---|---|
| Lowest Score | 70.0 |
| Lower Quartile | 88.5 |
| Median | 97.0 |
| Upper Quartile | 102.0 |
| Highest Score | 105.0 |

**29.6** 96 boys and 90 girls took a test.

The table shows information about the mean marks.

| | Boys | Girls |
|---|---|---|
| Number of students | 96 | 90 |
| Mean mark | 67.5 | 71.2 |

The pass mark for the test was 70.

Was the mean mark for all these students greater than the pass mark?

You must show your working.

### Questions referring to previous content

**29.7** One bag of grass seed covers an area of 3.66 m² and costs £4.99.
Fabio needs grass seed for a lawn of 32 m².

How much will the grass seed cost Fabio?
Give your answer to the nearest pound **[3 marks]**

 **29.8** The sum of the interior angles in an irregular polygon is 720°

The smallest angle is 20° and the largest is 220°.

The angles form an arithmetic sequence.

Work out the other four angles. **[3 marks]**

# Knowledge

## 30 Charts and graphs

### Bar charts

Bar charts are used for qualitative (categorical) data.

In a **bar chart**, the height of each bar shows the frequency. A **vertical line chart** uses lines instead of bars.

When drawing a bar chart

- label both axes
- make sure bars are the same width
- leave an equal gap between the bars.

**Dual bar charts** show two sets of data on the same bar chart.

**Worked example**

The dual bar chart shows the average number of different drinks sold per day at a café. What is the most popular drink on weekdays?

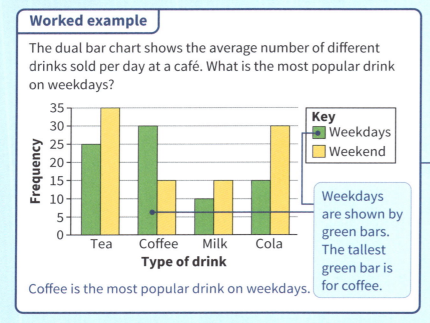

**Key**
- Weekdays
- Weekend

Weekdays are shown by green bars. The tallest green bar is for coffee.

Coffee is the most popular drink on weekdays.

The same data can be shown on a **composite bar chart**.

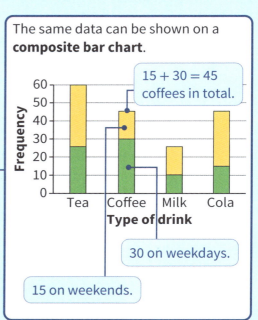

$15 + 30 = 45$ coffees in total.

30 on weekdays.

15 on weekends.

### Pie charts

A **pie chart** shows categories as a proportion of the total.

The frequency or amount of something is shown as a sector of a circle.

The angle of the sector shows the proportion of that category in relation to the total.

angle of sector = proportion (as a fraction) × 360

**Colour of 25 cars parked on a street**

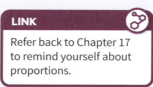

**LINK**

Refer back to Chapter 17 to remind yourself about proportions.

**WATCH OUT**

Be careful when comparing two pie charts with different totals!

## Pie charts

### Worked example

1. The vertical line chart shows the reasons for staff absence at a school.

**Reason for absence**

Draw a pie chart to show this data.

Total $= 4 + 6 + 2 = 12$ ● — Find the total number of days of absence.

Proportion $= \dfrac{4}{12} = \dfrac{1}{3}$ ● — Find the proportion of absence days caused by training.

Angle of sector $\dfrac{1}{3} \times 360$
$= 120°$

● — The angles of all sectors in a circle always add up to 360°.

● — Repeat to find the angles for 'Illness' and 'Other', and draw the pie chart.

**Key**
■ Training
■ Illness
■ Other

● — Remember to add a key.

2. A greengrocer has 81 pieces of fruit. He records the number of apples, bananas and pears.

   He draws a pie chart to show the results. Calculate how many apples there were.

   $\dfrac{a}{81} = \dfrac{160}{360} = \dfrac{4}{9}$ ● — Use equivalent fractions.

   $a = \dfrac{4 \times 81}{9} = 36$ apples

   $\dfrac{\text{number of apples, } a}{\text{total pieces of fruit}} = \dfrac{\text{angle for apples}}{\text{total angle in circle}}$

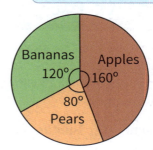

## Frequency polygons

**Frequency polygons** represent grouped data.

The frequency of each class is plotted against its midpoint, then each point is joined with straight lines.

## 30 Charts and graphs

# Frequency polygons

### Worked example

The grouped frequency table shows the heights of a sample of daisies.

| Height of daisies ($x$ cm) | Frequency |
|---|---|
| $2 < x \le 6$ | 2 |
| $6 < x \le 10$ | 8 |
| $10 < x \le 14$ | 24 |
| $14 < x \le 18$ | 4 |

Draw a frequency polygon to represent the data.

Plot each frequency against the midpoint of each class. For example, the midpoint of the first class is 4 cm. Remember to join the points with straight lines.

# Pictograms

**Pictograms** are used for **qualitative data**. They use images to represent the data. For example, you could represent eye colour with a pictogram that uses pictures of eyes.

| Eye colour | Number of children |
|---|---|
| Brown | ◉ ◉ ◉ ◉ |
| Hazel | ◉ ◉ ◉ |
| Blue | ◉ ◉ ◉ |
| Green | ◉ ◉ ◉ ◉ |

**Key**
◉ = 4 Children

Remember to add a key and use pictures that are simple to draw.

# Line graphs

**Line graphs** (or time-series graphs) show how data change over time. Time is always on the horizontal axis. For example, the average temperature for each month for a year.

## Scatter graphs

**Scatter graphs** show the relationship between two sets of data (or variables), such as the age and height of a group of children.

In this example, each point would represent one child.

Height of Children

**Correlation** describes how the data could be related.
It can range from strong to weak, and be positive or negative:

**Strong, positive**   **Weak positive**   **No correlation**   **Weak negative**   **Strong negative**

**Causation** means that one event causes another. Correlation between data sets does not necessarily mean that one event causes the other.

A **line of best fit** follows the trend of the data, ignoring **outliers**, and goes through (or near to) as many points as possible. The line of best fit should have roughly an equal number of data points on each side of it.
It does not need to go through the origin.

**Interpolation** means using the line of best fit to estimate a value. It is a good method for making predictions within the line of best fit.

**Extrapolation** is predicting results outside the data (that is, outside your line of best fit).

---

**Worked example**

The data shows the age of 10 chickens and the number of eggs each one laid over a month.

How many eggs would you expect a 2-year-old hen to lay?

| Age of the chicken | 1 | 2 | 2 | 3 | 3 | 4 | 4 | 4 | 5 | 5 |
|---|---|---|---|---|---|---|---|---|---|---|
| Number of eggs | 22 | 18 | 24 | 16 | 20 | 15 | 17 | 19 | 16 | 19 |

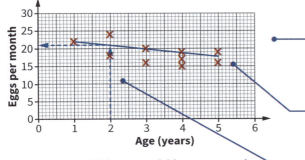

Plot a scatter graph for the information given. Take care of the scale and make sure points are plotted accurately.

Add a line of best fit by using your ruler to find a line through the middle of the points. Try to get the same number of points on each side. Ignore any outliers.

A 2-year-old hen would be expected to lay 21 eggs.

Draw a line up from 2 on the $x$-axis to the line of best fit, then across to the $y$-axis.

---

## 30 Charts and graphs

Learn the answers to the questions below, then cover the answers column with a piece of paper and write as many as you can. Check and repeat.

| | Questions | Answers |
|---|---|---|
| 1 | What is correlation? | There is a relationship between two sets of data. |
| 2 | What is a line of best fit? | A line which follows the trend of the data. |
| 3 | What are the types of correlation? | Positive correlation and negative correlation. |
| 4 | What type of data can you represent using a pictogram? | Qualitative. |
| 5 | What is extrapolation? | Making predictions outside the range of data. |
| 6 | Why do we need to be careful when extrapolating data? | It is not reliable because the relationship between the two variables might not hold for values outside the data. |
| 7 | What is the formula that is used to calculate the number of items in a pie chart sector? | number of items = (angle of sector ÷ 360) × (total number of items) |
| 8 | When drawing a frequency polygon do you use the lower bound, upper bound, or midpoint of the class to plot the points? | The midpoint. |

*Put paper here*

## Previous questions

Now go back and use these questions to check your knowledge of previous topics.

| | Questions | Answers |
|---|---|---|
| 1 | In algebra, what is a term? Give some examples. | A single number or variable, or numbers and variables multiplied together e.g. 2, $x$, $5y$ are all examples of terms. |
| 2 | Where is the line of symmetry on a quadratic graph? | Multiply or divide the previous term by a value. |
| 3 | When we divide 2 negative numbers what will be the result? | On the vertical line that passes through the turning point. |
| 4 | How do you convert cm² into mm²? | Multiply by $10^2 = 100$. |
| 5 | What does magnitude mean? | Magnitude means size. |

*Put paper here*

## Exam-style questions

**30.1** The table shows the monthly profits for a shop over a period of months.

| Month | Jan | Feb | Mar | Apr | May | Jun | Jul |
|---|---|---|---|---|---|---|---|
| Profit (£) | 3000 | 3500 | 4200 | 4500 | 4000 | 5000 | 6000 |

**(a)** Draw a time series graph to represent this data. **[2 marks]**

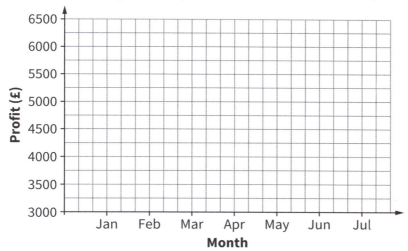

**(b)** Describe the general trend of the data. **[1 mark]**

**(c)** What might be misleading about this graph? **[1 mark]**

**30.2** The time series graph shows the attendance figures at a football ground over an eight-week period.

**(a)** Describe the general trend of the data. **[1 mark]**

**(b)** In which two weeks was attendance at its highest? **[1 mark]**

**(c)** Use the graph to estimate the attendance in Week 9. **[1 mark]**

**30.3** The scatter graph shows the relationship between the uphill gradient of some hills and the speed of some cyclists .

**(a)** Give a reason why you might ignore an outlier when drawing a line of best fit.                                                                                 **[1 mark]**

**(b)** What kind of correlation does the graph show?                         **[1 mark]**

**(c)** Ignoring the outlier, draw a line of best fit on the scatter graph. **[1 mark]**

**(d)** A cyclist is travelling at 15 km/h.

Use your line of best fit to estimate the uphill gradient.        **[1 mark]**

**(e)** By how many km/h would you expect the speed of a cyclist to decrease for every 1% increase in the gradient of a hill?    **[2 marks]**

**(f)** Marta uses a line of best fit to estimate that a cyclist travelling up a hill with 40% gradient should be travelling at a speed of 5 km/h.

Comment on Marta's assumption.                                      **[1 mark]**

**30.4** A scatter graph shows that there is positive correlation between the number of sandcastles made at a beach and the number of electric fans sold in a shop.

Trevor says that the increase in the number of sandcastles causes an increase in the number of sales of electric fans.

Give a reason why Trevor is wrong.                                      **[1 mark]**

**30.5** Linda planted 400 flower bulbs.

She planted a mix of daffodil, tulip and hyacinth bulbs.

The incomplete table and pie chart show some information about these bulbs.

| Type of bulb | Number planted |
|---|---|
| Daffodil | 180 |
| Tulip | |
| Hyacinth | |
| Total | 400 |

Complete the table and pie chart.                                      **[3 marks]**

**30.6** The frequency polygon shows the masses of some puppies.

(a) Use the frequency polygon to complete
the grouped frequency table. **[2 marks]**

| Mass $m$ (kg) | Frequency |
|---|---|
| $0 < m \le$ | |
| $< m \le 10$ | 12 |
| $< m \le$ | 6 |
| $< m \le 20$ | |
| | 3 |

Mass, m (kg)

(b) Estimate the mean mass of a puppy to the nearest kg. **[3 marks]**

**30.7** Anastasia asks each of her 30 school friends how many of their school lessons
they have enjoyed this week. The chart shows her results.

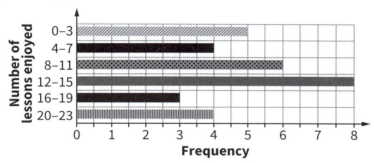

(a) What percentage of Anastasia's friends enjoyed more
than 7 lessons? **[2 marks]**

(b) Estimate the mean number of lessons that Anastasia's
friends enjoyed. **[4 marks]**

| Number of lessons | Frequency |
|---|---|
| 0–3 | |
| 4–7 | |
| 8–11 | |
| 12–15 | |
| 16–19 | |
| 20–23 | |

**Question referring to previous content**

**30.8** Euler's polyhedral formula states that, in any polyhedron, the number of faces
plus the number of vertices equals the number of edges plus 2.

Written mathematically: $F + V = E + 2$

A 3D shape has 10 vertices and 15 edges. Use the formula to find the number of
faces and then draw the shape. **[3 marks]**

## 31 Data collection, cumulative frequency, and box plots

### Types of data

| | Definition | Example |
|---|---|---|
| Data | Any information you can collect. | The heights of people who live in London. |
| Population | The entire group of things that you can choose from. | All the people who live in London. |
| Sample | A smaller set within the population that you can collect data from. | One person from every street in London. |
| Biased sample | A sample that doesn't represent the population well. | A group of men from a basketball team. |

In a **random sample**, each member of the population is equally likely to be chosen.

Data from a sample can be used to estimate properties of the whole population.

The larger the sample the more accurate the estimates.

> **Worked example**
>
> Jessica wants to know how many students in her school read per month.
>
> There are 1000 students in her school, so she wants to use a sample.
>
> She asks students in the library how many books they have read that month.
>
> **(a)** Jessica asks 10 students in total, and finds that two of them have read more than five books that month.
> Use this data to estimate how many students in the whole school have read more than five books that month.
>
> $\frac{2}{10} = \frac{1}{5}$     Write the proportion of the sample who have read more than five books as a fraction, and simplify it.
> of the students have read more than five books.
>
> $\frac{1}{5} \times 1000$     Multiply the proportion in the sample by the number in the population to find the estimate.
> $= 200$ students
>
> **(b)** Give **two** reasons why your estimate for part **(a)** may not be accurate.
>
> Any two from:
>
> - The sample is not random (not all students have an equal chance of being chose).
> - The sample may be biased (students in the library might be more likely to like reading than other students).
> - The sample size of 10 students is too small to be representative of the whole school.

### Capture-recapture

The **capture-recapture** method is a way to estimate the size of a population of animals.

When using the capture-recapture method:

- it is important to allow time between samples for the marked and unmarked animals to mix (but not too long as you don't want to have to consider the birth rate or death rate)
- you need to be confident that the animals can't escape the environment.

## Capture-recapture

To estimate the size of an animal population ($N$):

$$N = \frac{nK}{k}$$

Where:

$n$ = size of sample

$K$ = animals in the first sample that are marked

$k$ = marked animals found in the second sample

### Worked example

Elijah wants to estimate the population of rabbits in a fenced area.

He captures a sample of five rabbits, marks each of them and then releases them.

The next day he recaptures a second sample of eight rabbits and finds two of them are marked.

**1.** Estimate the number of rabbits in the area.

$$N = \frac{nK}{k}$$

$K = 5, k = 2$ and $n = 8$, so

$$N = \frac{8 \times 5}{2}$$

$$= 20$$

> Identify the information you have and substitute the values into the formula to estimate animal population size.

**2.** What assumptions have you made?

- the samples are random
- the marked and unmarked rabbits mix up
- the population hasn't changed overnight

> Three possible assumptions are listed, but you only need to give two.

**REVISION TIP**

You don't have to remember the formula – you can assume that the proportion of marked animals in the recaptured sample is the same as the proportion in the whole population. Here, the proportion of marked rabbits is $\frac{2}{8}$. Assuming that $\frac{2}{8}$ of all the rabbits in the area have been marked, the total number of rabbits in the area is $5 \times 4 = 20$.

## Cumulative frequency graphs

**Cumulative frequency graphs** show the total frequency up to each point.

### Worked example

Here is some information about the heights of some horses.

**(a)** Draw a cumulative frequency graph to represent this information.

> First, add a cumulative frequency column to the table.

| Height, $h$ (m) | Frequency | Cumulative frequency |
|---|---|---|
| $2.1 < h \leq 2.2$ | 2 | 3 |
| $2.2 < h \leq 2.3$ | 3 | 5 |
| $2.3 < h \leq 2.4$ | 7 | 12 |
| $2.4 < h \leq 2.5$ | 7 | 19 |
| $2.5 < h \leq 2.6$ | 1 | 20 |

## 31 Data collection, cumulative frequency, and box plots

### Cumulative frequency graphs

**Worked example**

Plot a graph with *Cumulative frequency* on the *y*-axis, and the end-point of each height class on the *x*-axis. Join the points with a smooth curve.

**(b)** Estimate the median height.

Median ≈ 2.37 *m*

There are 20 horses in total, so for the median, draw a line from 10 on the *y*-axis, across to the graph line, then down to the *x*-axis.

**(c)** Estimate the interquartile range.

LQ ≈ 2.30 *m* : UQ ≈ 2.44 *m*

Interquartile range = 2.44 − 2.30

= 0.14 m

For the *LQ* and *UQ*, draw lines from 5 and 15 on the *y*-axis across to the graph, and read the value on the *x*-axis.

**(d)** Estimate the percentage of horses with height less than 2.4 m.

12 horses, which is 60% of 20

Draw a line up from 2.4 m on the *x*-axis.

Then calculate the number of horses as a percentage of the total number of horses.

### Box plots

A **box plot** can be used to show the median, range and interquartile range of data and is useful for comparing sets of data.

To identify the quartile values:

Lower quartile = $\frac{1}{4}$ (number of values in set + 1)

Upper quartile = $\frac{3}{4}$ (number of values + 1)

**Features of a box plot:**

**Key terms**

Make sure you can write a definition for these key terms

biased sample   box plot   cumulative frequency graph
frequency density   histogram   capture-recapture
population   random sample

## Box plots

### Worked example

The heights of a group of children are (in cm):

144  125  141  135  136  132  142  160  133

> List the heights in order from smallest to largest.

**(a)** Identify any outliers in the data.

An outlier is any value that more than $1.5 \times \text{IQR}$ above the upper quartile or $1.5 \times \text{IQR}$ below the lower quartile.

125, 132, 133, 135, 136, 141, 142, 144, 160

There are 9 children:

Median = 5th value = 136 cm

> Identify the median.

Lower quartile:

$\frac{1}{4}((9+1)) = 2.5$

mean of 2nd and 3rd values = 132.5 cm

Upper quartile

$\frac{3}{4}(9+1) = 7.5$

> Find the quartiles.

mean of 7th and 8th values = 143 cm

IQR = 143 − 132.5 = 10.5 cm

> Calculate the IQR

$1.5 \times \text{IQR} = 1.5 \times 10.5$

= 15.75

> Calculate $1.5 \times \text{IQR}$

Lower quartile − $1.5 \times \text{IQR}$ = 132.5 − 15.75

= 116.75

Upper quartile + $1.5 \times \text{IQR}$ = 143 + 15.75

> Calculate the values below and above which any outliers would lie.

= 158.75

125, 132, 133, 135, 136, 141, 142, 144, (160)

> 160 cm lies above 158.75 cm so is an outlier.

**(b)** Draw a box plot to represent these heights.

> Use the values to draw the box plot.

## Histograms

**Histograms** can be used to represent grouped data. They are particularly useful when the class widths are different.

In a histogram:

- the *y*-axis shows the **frequency density**

$$\text{frequency density} = \frac{\text{frequency}}{\text{class width}}$$

- the **area** of a bar is proportional to the frequency of the class.

## Histograms

### Worked example

Here is some information about the heights of 100 Year 10 students.

| Height, $x$ (cm) | Frequency | Frequency density |
|---|---|---|
| $100 < x \leq 110$ | 13 | 1.3 |
| $110 < x \leq 135$ | 24 | 0.96 |
| $135 < x \leq 150$ | 36 | 2.4 |
| $150 < x \leq 160$ | 17 | 1.7 |
| $160 < x \leq 200$ | 10 | 0.25 |

**(a)** Draw a histogram to represent this information.

> Start by calculating the frequency density.

### Worked example

**(b)** Work out an estimate of the median.

Give your answer to 1 d.p.

Median value = 50th value

Median is in the interval $135 < x \leq 150$

Median value = $50 - 13 - 24$

   = 13th value in the interval

$\frac{13}{36} = \frac{x}{15}$, so $x = 5.4$

Median value = $135 + 5.4 = 140.4$

> Median value = middle of the data set $(100 \div 2)$.

> Find the interval containing the median by adding the frequencies in the table.

> Find the position of the median in this interval by subtracting the frequencies of the previous intervals.

> Find the 13th value. Assume that the group is equally spaced, so all 36 values are the same distance from each other.

### Worked example

Some children took a short test.

The histogram shows information about the times taken.

Three children completed the test in less than 6 seconds.

Work out the number of children that took between 6 and 10 seconds to complete the test.

3 children completed the test in under 6 s, so for the 0-6 s class,

frequency density (f.d.) = $\dfrac{\text{frequency (f)}}{\text{class width (c.w.)}} = \dfrac{3}{6} = 0.5$

So the frequency density axis has a scale of 1 square = 0.5

Hence the number of children that took between 6 and 10 seconds = $2.5 \times 4 = 10$ children

## 31 Data collection, cumulative frequency, box plots

Learn the answers to the questions below, then cover the answers column with a piece of paper and write as many as you can. Check and repeat.

| | Questions | Answers |
|---|---|---|
| 1 | Why should you not use a biased sample? | It will not represent the population well. |
| 2 | Why do we use the capture recapture method? | To estimate the size of a population. |
| 3 | What is a cumulative frequency graph? | A graph that shows the total frequency up to each point and can be used to estimate the median and quartiles. |
| 4 | What do you plot on the $x$-axis of a cumulative frequency graph from grouped data? | The end point of each class interval. |
| 5 | How do you estimate the median? | Draw a line across from the $y$-axis at 50% of the total frequency, and then down to the $x$-axis. This value is the median. |
| 6 | How do you estimate the upper and lower quartiles from a cumulative frequency graph? | Draw lines across from the $y$-axis at 25% and 75% of the total frequency, and then down to the $x$-axis. |
| 7 | What is a box plot used for? | For comparing sets of data. |
| 8 | What type of data is displayed on a histogram? | Grouped data. |

*Put paper here*

## Previous questions

Now go back and use these questions to check your knowledge of previous topics.

| | Questions | Answers |
|---|---|---|
| 1 | If the ratio $a:b$ is $10:7$, what is the ratio of $b:a$? | $7:10$ |
| 2 | What are the conditions for a triangle to be congruent? | SSS, SAS, ASA, RHS |
| 3 | What is the cosine rule to find an angle? | $\cos A = \dfrac{b^2 + c^2 - a^2}{2bc}$ |
| 4 | What is the exact value of: sin 60? | $\dfrac{\sqrt{2}}{3}$ |
| 5 | What is the formula used to find the area of a triangle using trignometry? | $A = \dfrac{1}{2}\, ab \sin C$ |

*Put paper here*

# Practice

## Exam-style questions

**31.1** Sam wants to predict the football results for future matches.
She uses a list of results for matches that took place five years ago.

Explain why the data might not be reliable. **[1 mark]**

**31.2** 180 sixth form students want to celebrate their end-of-course
exam results. A sample of 40 students is taken to decide how
they should celebrate. Their preferences are shown in the table.

| Type of celebration | Frequency |
|---|---|
| Cinema trip | 11 |
| Party | 10 |
| Meal out | 11 |
| Day at funfair | 8 |

    **(a)**    Work out how many of the 180 students you think would
like a day at the funfair. **[2 marks]**

    **(b)**    State one assumption that you have made in your answer
to part **a**. **[1 mark]**

**31.3** **(a)**    Write what is meant by a random sample. **[1 mark]**

Harriet wants to find the favourite foods of students in her school.

She gives five of her best friends a questionnaire.

    **(b)**    Describe

        **(i)**    the population **[1 mark]**

        **(ii)**    the sample. **[1 mark]**

    **(c)**    Give a reason why her sample might be biased. **[1 mark]**

    **(d)**    Write two ways that her sample can be improved. **[2 marks]**

**31.4** A researcher wants to estimate the the approximate number of
gulls at a seaside resort.

On Tuesday they catch 50 gulls.

They put a tag on each gull and let it go.

On Wednesday they catch 234 gulls.

18 of these gulls have a tag.

    **(a)**    Work out an estimate for the number of gulls at the resort. **[3 marks]**

    **(b)**    State one assumption you have made in your answer to part **a**. **[1 mark]**

The researcher thinks that the tags on some of the gulls fell off after
they were tagged.

    **(c)**    Describe what effect this would have on your answer to part **a**.
Give a reason for your answer. **[1 mark]**

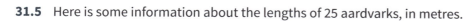 

**31.5** Here is some information about the lengths of 25 aardvarks, in metres.

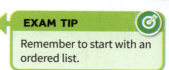

| 1.3 | 1.6 | 1.8 | 1.9 | 1.1 | 1.4 | 1.5 | 1.0 | 0.8 | 0.8 | 1.5 | 1.7 |
|-----|-----|-----|-----|-----|-----|-----|-----|-----|-----|-----|-----|
| 1.7 | 1.9 | 1.4 | 1.1 | 0.8 | 2.0 | 1.7 | 2.0 | 1.3 | 1.4 | 1.8 | 0.9 | 1.1 |

Draw a box plot to represent this information. **[4 marks]**

 **31.6** Here is some information about the amount of time year 10 students spend on maths homework.

The times are in minutes.

| Minimum | Interquartile range | Maximum | Lower quartile | Median |
|---------|---------------------|---------|----------------|--------|
| 22 | 8 | 48 | 32 | 34 |

**(a)** Draw a box plot to represent this information. **[2 marks]**

This box plot represents the the amount of time year 11 students spend on maths homework.

**(b)** Compare the distribution of the year 11 times to the year 10 times. **[3 marks]**

**31.7** 200 students get ready for school.

Here is some information about the times taken.

**(a)** On the grid, draw a cumulative frequency graph. **[3 marks]**

| Time ($t$ minutes) | Frequency |
|--------------------|-----------|
| $0 < t \le 5$ | 30 |
| $5 < t \le 10$ | 74 |
| $10 < t \le 15$ | 56 |
| $15 < t \le 20$ | 22 |
| $20 < t \le 25$ | 18 |

**(b)** Work out the median time. **[1 mark]**

**(c)** Neve says, 'More than 30% of pupils take less than 8 minutes to get ready for school.'

Is Neve correct?

Show working to support your answer. **[3 marks]**

# Exam-style questions

**31.8** The frequency table shows the speeds, in km/h, of some cars on a motorway.

**(a)** Complete the missing information in the table and in the histogram. **[4 marks]**

**EXAM TIP**

Frequency Density

$= \dfrac{\text{Frequency}}{\text{Class Width}}$

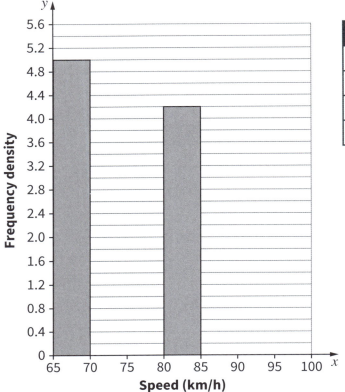

| Speed ($s$ km/h) | Frequency |
|---|---|
| $65 < s \le 70$ | |
| $70 < s \le 80$ | 44 |
| $80 < s \le 85$ | |
| $85 < s \le 100$ | 18 |

**(b)** Estimate the mean speed. **[3 marks]**

**(c)** Estimate the median speed, stating any assumptions you have made. **[3 marks]**

**EXAM TIP**

Find the class that contains the median, and then use a proportional method.

**31.9** Here is some information about the masses of some dogs.

**(a)** Nanveet says, 'The mean might not be the best average to use for this data.'

Give a reason to support Nanveet's claim. **[1 mark]**

**(b)** Work out the class interval that contains the median. **[1 mark]**

**(c)** On the grid on the next page, draw a frequency polygon for the information in the table. **[2 marks]**

| Mass ($m$ kg) | Frequency |
|---|---|
| $0 < m \le 4$ | 3 |
| $4 < m \le 8$ | 6 |
| $8 < m \le 12$ | 5 |
| $12 < m \le 16$ | 8 |
| $16 < m \le 20$ | 20 |

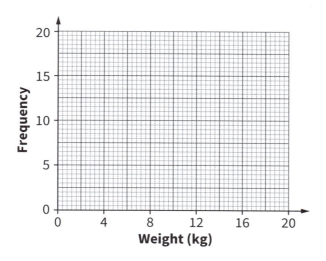

**31.10** The cumulative frequency graph shows information about Test 1 marks for 32 pupils.

The lowest mark was 10 and the highest mark was 48.

**(a)** On the grid, draw a box plot to represent this information. **[3 marks]**

The box plot shows the information about the same pupils' marks for Test 2.

**(b)** Cain says, 'The median score for the second test is higher, so that test must have been easier.'

Is Cain correct?

Show working to support for your answer. **[2 marks]**

**Questions referring to previous content**

**31.11** Each of these equations matches to one of the graphs.

$$y = \frac{1}{x} \qquad y = x^2 \qquad y = x^3 \qquad y = x$$

Write the correct equation below each graph. **[2 marks]**

v

Great Clarendon Street, Oxford, OX2 6DP, United Kingdom

Oxford University Press is a department of the University of Oxford. It furthers the University's objective of excellence in research, scholarship, and education by publishing worldwide. Oxford is a registered trade mark of Oxford University Press in the UK and in certain other countries.

Written by Naomi Bartholomew-Millar, Paul Hunt and Victoria Trumper

The moral rights of the authors have been asserted

The publisher would also like to thank Katie Wood and Jemma Sherwood for their work on the first edition of AQA GCSE Maths Higher Revision Guide (978-138-200647-7) and Revision Workbook (978-138-200648-4) on which this revision guide is based.

First published in 2023

British Library Cataloguing in Publication Data
Data available

978-138-203983-3

10 9 8 7 6 5 4 3 2 1

The manufacturing process conforms to the environmental regulations of the country of origin.

Printed in the UK by Bell and Bain Ltd, Glasgow

**Acknowledgements**

**Artworks:** QBS Learning

The publisher would also like to thank Deb Friis and Katherine Pate for sharing their expertise and feedback in the development of this resource.

Although we have made every effort to trace and contact all copyright holders before publication this has not been possible in all cases. If notified, the publisher will rectify any errors or omissions at the earliest opportunity.

Links to third party websites are provided by Oxford in good faith and for information only. Oxford disclaims any responsibility for the materials contained in any third party website referenced in this work.